T0013288

GENETICS 101

GENETICS 101

FROM CHROMOSOMES AND THE DOUBLE HELIX TO CLONING
AND DNA TESTS, EVERYTHING YOU NEED TO KNOW ABOUT GENES

BETH SKWARECKI

Adams Media

New York London Toronto Sydney New Delhi

Adams Media
An Imprint of Simon & Schuster, Inc.
100 Technology Center Drive
Stoughton, MA 02072

First Adams Media hardcover edition July 2018

ADAMS MEDIA and colophon are trademarks of Simon & Schuster.

For information about special discounts for bulk purchases, please contact Simon & Schuster Special Sales at 1-866-506-1949 or business@simonandschuster.com.

The Simon & Schuster Speakers Bureau can bring authors to your live event. For more information or to book an event contact the Simon & Schuster Speakers Bureau at 1-866-248-3049 or visit our website at www.simonspeakers.com.

Manufactured in Italy

10 9 8 7 6

Library of Congress Cataloging-in-Publication Data
Skwarecki, Beth, author.
Genetics 101 / Beth Skwarecki.
Avon, Massachusetts: Adams Media, 2018.
Series: Adams 101.
Includes index.
LCCN 2018011664 (print) | LCCN 2018012734 (ebook) | ISBN 9781507207642 (hc) | ISBN 9781507207659 (ebook)
LCSH: Genetics--Popular works. | Science--Popular works. | BISAC: SCIENCE / Life Sciences / Genetics & Genomics. | SCIENCE / Life Sciences / Human Anatomy & Physiology. | SCIENCE / General.
LCC QH437 (ebook) | LCC QH437 .S59 2018 (print) | DDC 572.8--dc23
LC record available at https://lccn.loc.gov/2018011664

ISBN 978-1-5072-0764-2
ISBN 978-1-5072-0765-9 (ebook)

CONTENTS

INTRODUCTION

Less than two centuries ago, all people really knew about genetics was that children tend to look like their parents and that careful breeding of dogs or horses or crops can result in bigger and better dogs or horses or crops. We've learned a lot since then.

In the 1800s, a monk named Gregor Mendel figured out that traits of pea plants—like whether peas were yellow or green—were passed down from parent to child in a way that could sometimes hide traits so they appeared to skip a generation. He figured out how to predict whether and when a hidden trait would show up next.

Around the same time, naturalist Charles Darwin figured out that species evolve over time. The traits of pets and crops are influenced by a farmer who breeds them, but according to Darwin's theory of evolution by natural selection, it is nature, rather than human judgment, that determines which creatures live long enough to have offspring. Darwin knew the whole idea hinged on some mysterious way that parents can pass down traits to their children, but he had no idea how that might work.

And then, in the 1950s, Rosalind Franklin managed to form DNA into a crystal and take an x-ray photograph that revealed its structure. James Watson and Francis Crick built on her work to deduce that the DNA molecule had the shape of a double helix and that DNA's structure was uniquely suited to pass down traits from one generation to the next. Over the remaining decades, scientists have worked out the details of exactly how DNA makes us who we are—and how we can tinker with it.

This book will explain genetics, which is the study of how living things give their offspring the instructions, or genes, for particular traits. We'll also talk about genomics, which is a related field that studies the totality of all the information contained in your DNA. Along the way, we'll cover other bits of biology as needed. We'll do all this with a focus on you and what's going on in your body, plus a few things you might see in the news.

Along the way, we'll take some detours to visit the genomes of animals, plants, bacteria, and even viruses. You have more in common with all of these creatures than you probably realize.

First, we'll learn about the nuts and bolts of deoxyribonucleic acid—DNA—itself. It's a stringy substance that, on a smaller-than-microscopic scale, is an incredibly long molecule. You have forty-six of these strands stuffed into the nucleus of each cell in your body, and each strand contains instructions for building and maintaining every part of your body. These instructions are in a chemical language that we'll learn to decode.

We'll see how your cells read that code and carry out the instructions. Often the instructions tell the cells to build a protein, so we'll learn what these proteins do too. Some of them give your eyes and hair and skin their distinctive colors. Some help your body to process food and drugs. Some are so important to the way your body functions that if they aren't built in exactly the right way, you could end up with an increased risk for cancer or other health conditions.

We'll also learn about how your DNA got to you in the first place: how it was passed down from your parents and what it can tell you about your family tree. And we'll see what you can learn from personal genomics services that promise to reveal your deepest secrets based on a sample of saliva. Finally, we'll take a look at what scientists and companies are doing with DNA, from genetically modifying crops to improving treatments for cancer.

YOUR CELLS' INSTRUCTION MANUAL

What DNA Does

DNA is what makes us who we are. But how does it do that?

To answer that question, we have to zoom in to a level even smaller than what microscopes can see. DNA is a long, stringy molecule whose job is to carry information. To understand that, think for a minute about this book. It's just letters, one after another, that taken together form words and sentences and chapters. A DNA strand is made up of millions of chemical components that function like letters, spelling out an instruction manual with all the information it takes to build and run a human body. (Or an animal's body or even a plant or a bacterium. Every living thing has DNA.)

Noncoding DNA

There's more to our DNA than just recipes, though. Think of the genome as a deluxe cookbook with a ton of extra information, like how to plan a dinner party or suggested menus for a week's dinners. That information is helpful so you know when to make the recipes. But it's also a sloppy cookbook: there might be three versions of the same dish, and only one of them is worth making. Perhaps there are even some recipe cards and scraps of paper tucked into the pages, things that you're not quite sure where they came from but you're not sure if it's okay to throw them away. Our DNA has scraps like that too.

There is far more information in DNA than in a book, though. If you printed it out, our **genome**—all the information carried in our DNA—would fill twelve thousand books this size.

Our genome isn't just one string of DNA; it's actually split into pieces called **chromosomes**. I like to think of our twenty-three chromosomes as a recipe collection in twenty-three enormous volumes. Like a real cookbook, DNA contains short sets of instructions—think of them as recipes. Each recipe, or **gene**, contains the instructions to make one tiny piece of who we are.

Since we are all different, our recipes vary slightly. My genes include instructions on how to make a brown pigment and put it into my hair follicles. But your hair may be colored differently from mine if your genes encode a recipe for a different pigment. Or perhaps that page in your recipe book is blank, and you don't put any pigment in your hair at all.

You have two copies of this cookbook encyclopedia in each cell of your body. Cells are, in a sense, the kitchens where the recipes are made.

WHAT IS A CELL?

Your body contains over thirty-seven trillion cells. That's a huge number, right? You have more cells in your body than there are dollars in the national debt or stars in the Milky Way.

You have skin cells, muscle cells, fat cells, nerve cells, and bone cells, just to name a few. They're all so small you can only see them with a microscope. Every time you scratch an itch, hundreds of skin cells flake off, and you don't even notice.

Nearly every one of those cells contains all of the DNA we just talked about. Double that, actually, since you keep two copies around—the one you got from your mom and the one you got from your dad.

Mitochondria Have DNA Too

Most of our organelles are pretty boring, but there's a special one called the mito-chondrion (plural: mitochondria) that helps turn food into energy. It's so special that it has its own DNA that it doesn't share with us. Scientists think this is because mitochondria used to be free-roaming bacteria that one day got eaten—but not digested—by a larger cell. After millions of years, we're like best friends: inseparable.

Our cells have different compartments, or **organelles**, separated from each other by membranes. We keep those two full sets of our DNA in their own organelle called the **nucleus**. This way they're safe from all the chaos going on in the rest of the cell. (Think of it like a special library for our cookbook collection.)

CHOOSING RECIPES

If all of our cells have the same cookbook collection, why aren't they all following the same instructions all the time? If they did that, all thirty-seven trillion of our cells would look alike.

What actually happens is that skin cells only use the genes that are necessary to do skin cell things. Muscle cells only use the genes that help them do muscle cell things. (Skin cells and muscle cells have plenty of things in common, of course, so some recipes are used by both.)

Even in a single cell type, things change all the time. Brain cells use different genes during the day than at night, for example. Your stomach cells use different genes when you're digesting food than they do during those long stretches between meals. And you'll use a different mix of genes as an adult than when you were a baby.

ATOMS AND MOLECULES

The Building Blocks of DNA

DNA is a huge molecule, made of millions of atoms.

The best way to understand the difference between atoms and molecules is to sit down with a molecular model kit. You can sometimes find these in chemistry classrooms or college bookstores, which makes them seem very serious, but in reality, a molecular model kit is just a very fun toy.

Try Building Alcohol

If water is too boring, you can make ethanol, which is the kind of alcohol that's in beer and wine. Start with a carbon and add three hydrogens. On the fourth toothpick, stick another carbon, and give that carbon two hydrogens. The second carbon should now have three toothpicks in it, so for the fourth toothpick you'll add the hydroxyl group, which is just an oxygen atom that has a hydrogen attached. That -OH group is what makes it an alcohol.

If you don't have one, that's fine! You can play along at home with a bag of gumdrops and a box of toothpicks.

Let's start with a simple molecule: water. You probably know water's chemical formula already: H_2O. That means it has two hydrogen atoms and one oxygen atom. If you're doing the candy version of this exercise, grab a red gumdrop and stick two toothpicks into it. That red gumdrop is your oxygen atom. Take two white gumdrops to represent the two atoms of hydrogen and stick them at the other end of each toothpick. You've just made H_2O.

If you're lucky enough to have a model kit, the hydrogen atoms will be built with just one socket where a connector can fit in. The oxygen atoms will have two sockets. That's because in real life, oxygen can (normally) only make two bonds. Hydrogen makes just one.

The Atoms We'll Be Working With

Most of the molecules in our cells can be made with just six atoms. Think of them as your organic chemistry starter kit:

- Carbon (4 bonds)
- Hydrogen (1 bond)
- Oxygen (2 bonds)
- Nitrogen (3 bonds)
- Phosphorus (it's complicated)
- Sulfur (likewise)

  Of these, you only need the first five to build DNA.

Carbon, on the other hand, can make four bonds, so the little black spheres that represent carbon will have four sockets in them. That's the advantage of the model kit: each piece has an appropriate number of sockets. If you're using gumdrops, you have to remember, on your own, how many toothpicks to put into each atom.

WHAT ARE ATOMS?

Have you ever heard of the periodic table of elements? It's that weird-shaped chart with one square for each known element. Some are things you've heard of: carbon, hydrogen, and oxygen, for example. Others are metals, and you've heard of a lot of these too: gold, silver, aluminum, copper. Neon, the gas that fills the tubes in light-up neon signs, is also an element.

These elements are really just the flavors that atoms can come in. What determines the flavor of an atom? It's the number of protons the atom has. Hydrogen has one proton, helium has two, and so on. If you're wondering about some of the elements we've already met, carbon has six protons, nitrogen has seven, and oxygen has eight. Gold has seventy-nine, and uranium has ninety-two.

Protons have a positive charge, so the more protons an atom has, the more negatively charged electrons it can collect.

You don't need to understand protons and electrons (or their neutrally charged buddies, neutrons) to be able to understand this book, so if this sounds like too much chemistry all at once, don't sweat it. We just mention them because the electrons are what determine the number of bonds an atom can make.

GIANT MOLECULES

As you tinker with your gumdrops and toothpicks, you might get carried away and decide to make the biggest molecules you can. And if you have a mega-sized bag of gumdrops, you'll find that molecules can be enormous!

For example, say you look up how to make a molecule of glucose—it forms a ring, so it looks kind of like a spiky crown. Make a bunch of these, and you can start chaining them together to build starch, the carbohydrate that provides most of the calories in foods like bread, pasta, and rice. Molecules like starch that are made from repeats of smaller building blocks are called **polymers**.

DNA is another polymer, but it's a bit more complicated than starch. Instead of one building block that repeats over and over, DNA has four different types of building blocks. Its pieces also come together in a way that makes a unique structure called a **double helix**. We'll learn more in the next sections about how this molecule is put together.

BUILDING BLOCKS

We've done a lot of building today, and we're about to do some more. Here's your cheat sheet for what builds what:

- Atoms are the smallest possible piece of an element. They are the building blocks of molecules.
- Molecules are the smallest possible piece of a compound, such as water or DNA. (Imagine a glass of water; the smallest item in the glass would be a single H_2O molecule.)
- Glucose, a sugar, is the building block of starch.
- Amino acids are the building blocks of proteins.
- Nucleotides are the building blocks of DNA.

NUCLEOTIDES

An Alphabet of Four Letters

Now that you know how atoms can join together to form molecules, we're going to learn how they make DNA. This isn't just so you can build a giant gumdrop-and-toothpick DNA molecule for your next party (although that's a great idea, isn't it?). The **structure** of DNA is crucial to understanding how DNA can carry instructions for the cell. Its structure also determines how parents can copy its information to pass down to their children. So, if we want to understand genetics, we have to zoom in to the molecular level.

Molecular Structure

If you haven't studied chemistry, you probably think of chemicals as liquids. But zoom in to the individual molecules, and each one has a three-dimensional shape. Molecules can bump into each other. They can wrap around each other. They take up space. We call the three-dimensional shape of a molecule its **structure**.

Since a single strand of DNA is a polymer—made of repeating units—we can't build it until we have the right units to start with. Those repeating units are called **nucleotides**.

There are a few different kinds of nucleotides, but they all have three components:

- A **nitrogenous base**, which can come in one of four different versions, nicknamed A, T, G, and C. We'll learn more about these in a minute.

- A **sugar**, specifically a kind called deoxyribose. This is different from the table sugar you put in your coffee, but in a sense, it's in the same chemical family. Deoxyribose is in the shape of a five-sided ring, and it's connected to the nitrogenous base.
- A **phosphate**, which is a phosphorus atom surrounded by oxygen atoms. This is also attached to the sugar.

To attach one nucleotide to the next, you attach the new nucleotide's phosphate group to the sugar on the previous nucleotide.

In real life, in your cells, there's no pair of hands stringing gumdrops together. Instead, when it's time to make more DNA, special proteins pick up nucleotides that are floating around the cell. They stick them onto the bottom of the growing DNA chain.

ATP

The structure of a nucleotide—base, sugar, phosphate—is something you might recognize if you've studied biology before. ATP, or adenosine triphosphate, has the same structure. ATP is best known as a source of energy for the cell, thanks to the high-energy bonds between its three phosphate groups. The nucleotides that can be incorporated into DNA are built the same way: when they're floating around the cell, they usually have three phosphate groups. It's like carrying their own little batteries to provide the energy needed to attach themselves to the growing DNA strand.

(From here on out, we're going to deal with gumdrops by the millions, so it may be best to stick with imaginary gumdrops rather than the real thing.)

THE FOUR NITROGENOUS BASES

Just like we spell words with the twenty-six letters of our alphabet, the information in DNA is "spelled" in an alphabet of four chemicals. With the right lab equipment, you can read down the strand of DNA and see what it says. Perhaps something like:

...ATCGTCTGACTGACGACTGATCGTAGTCGATCGATGCG-TACGATGCGTA...

Each of those four letters—A, T, G, or C—represents a different kind of nucleotide. The differences are in the part of the nucleotide called the nitrogenous base. The bases' full names are:

- Adenine
- Thymine
- Guanine
- Cytosine

They are called nitrogenous bases because they are basic—the opposite of acidic—and they contain a lot of nitrogen atoms in their structure. Thymine and cytosine are in the shape of flat rings, while adenine and guanine have a structure with two rings, like a figure eight.

THE DOUBLE HELIX

A Spiral Staircase

By this point in the history of the world, probably everyone reading this book has seen a picture meant to represent DNA. Any company whose work has to do with genetics or biotech has a doodle of DNA in their logo, for example. Any video that mentions genes or cells will include animations of a lumpy, oddly lit DNA strand. But these depictions aren't always accurate, so let's take a look at how DNA is really built.

DNA's True Shape

Real DNA has a wonky, asymmetrical look to it. Imagine that you have a ladder—that's your DNA—but before twisting it into a helix, you also pull the two uprights toward each other, bending the rungs. In real DNA, one side of the ladder (where the uprights are closer together) is called the minor groove, and the other side is the major groove. When you twist it, the spiral won't look neat and symmetrical. DNA, like life, is kind of messy.

BASE PAIRING

Those nitrogenous bases we discussed in the last section are responsible for a very important feature of DNA. They like to stick to other nitrogenous bases, and each has a specific partner.

Take adenine, for example. It's shaped like two flat rings, and as part of DNA it's attached to the sugar-phosphate backbone. But on the side that isn't connected to the backbone, it has some atoms that can make **hydrogen bonds**.

Remember when we were playing with gumdrops and tooth-picks? The toothpicks represented **covalent bonds**, which are chemical bonds that keep two atoms pretty tightly connected. Hydrogen bonds aren't like that. They're a less permanent sort of link between two atoms or molecules, resulting from the atoms' electrical charge. It may help to think of a hydrogen bond as being like the static electricity that can keep a sock from the dryer stuck to your pant leg. You can pull it off any time you like, since it's not permanently connected to your pants.

Adenine and thymine are shaped in such a way that they can make hydrogen bonds with each other. (Adenine has a hydrogen that wants to stick to thymine's oxygen, and thymine has a hydrogen that wants to stick to adenine's nitrogen.) The other two bases, cytosine and guanine, also make a pair, and they actually manage three hydrogen bonds rather than two.

Best Buddies

Here's how the base pairings work:

- Adenine pairs with thymine.
- Guanine pairs with cytosine.

These pairings are specific: if you bring two adenines next to each other, or an adenine and a cytosine, they won't hydrogen-bond very well. But bring an adenine near a thymine, or a cytosine near a guanine, and they'll stick together like socks out of the dryer.

TWO STRANDS

So far, we've described one strand of DNA: a sugar-phosphate backbone with nitrogenous bases just dangling free. But in our cells, DNA doesn't hang around all by itself. It's typically paired with a **complementary strand**.

That's complementary in the sense that the second strand is a perfect counterpart, or complement, to the first one. If one strand is full of adenines, its partner will be full of thymines. In reality, DNA strands have complex, or sometimes random, sequences of nitrogenous bases. The second strand is a reflection of that. If one strand has CTAGGC, for example, the other has to be GATCCG.

Watch Your Spacing

Remember how some of the bases have a single ring shape, and others have a double ring? It just so happens that each matching pair includes a single and a double. That means the backbones of DNA are always the same distance (three rings) apart.

HOW DNA MAKES ITS MATCH

The cell doesn't create two separate strands and then join them together. Remember how the nitrogenous bases want to stick to other nitrogenous bases? It would be like what happens when a toddler finds a roll of masking tape: you'd get a huge strand stuck to itself in so many places you'd have to throw it out and start over.

Instead, our cells have a very clever way of creating the second strand of DNA. First, let's think about when and where this happens.

When one cell needs to divide into two—say you're growing more skin cells to help heal a wound—each cell needs its own copy of the DNA.

The cell's machinery pries open the two strands, so there is a short section of DNA where the bases have nothing to stick to. Fortunately, the liquid inside this part of the cell contains a soup of nucleotides. These "free" nucleotides find their matches on the newly exposed parts of DNA. Special proteins called enzymes link the nucleotides to each other, creating two new DNA double strands. We'll learn more about this process in the section titled "DNA Replication."

THE END RESULT

The bases pair with each other, sticking the two strands of DNA together. And because the nucleotides have a somewhat curvy shape, they don't stack up straight like a ladder; instead they twist, like a spiral staircase.

The resulting shape is called a **double helix**: "double" because it's made of two strands, and "helix" from the Greek word for a spiral.

There's one more quirk that you should know: one of the two strands is upside down.

Technically there's no "up" and "down" in a cell, but the strands do have a direction to them. We usually draw a strand of DNA so the phosphate is on top, with the sugar below. Biochemists like to number the carbon atoms in the sugar, from one to five. The phosphate is attached to the fifth carbon atom, so we call the phosphate end of the DNA strand the **5' end**, pronounced "five prime end." The other end of the strand, named after the sugar's other attachment point, is the **3' end**, or "three prime end." Whenever we're adding to a DNA strand, the new nucleotides are added to the 3' end.

HOW ALL THAT DNA FITS INTO CELLS

How Big Is DNA, Anyway?

DNA is big and small at the same time. It's a molecule, so it's very, very tiny. You can't see it with a regular microscope, no matter how powerful the lenses. On the other hand, strands of DNA are very, very long.

Measurements of DNA

Here are the important measurements of a strand of DNA:

- **2 nanometers:** the distance from one backbone to another (the width of a "rung" of the ladder).
- **3.6 nanometers:** the length of DNA that it takes to make one full turn of the spiral.
- **0.34 nanometers:** the vertical space between base pairs (between rungs of the ladder). That means a full turn of the spiral includes about ten rungs.

We've already seen that the information in one copy of human DNA is enough to print twelve thousand books the size of this one. Now let's consider what that means in terms of base pairs. We have three billion base pairs in our DNA. If that were all in one long molecule (which it's not; it's divided into twenty-three), we would have a nearly invisible rope about a meter long, or just over three feet.

We actually have two copies of all this information. And when we say that, we're not referring to the two strands of the double helix! Those two strands match each other perfectly, so they don't really count as separate copies of the information. Instead, for each double helix you've got, you also have another complete double helix with almost the same information. You get one from each of your biological parents, so they will differ slightly. If that's confusing, don't worry—we'll learn more in a future section about how parents pass DNA to their children.

We actually have enough DNA to stretch out more than six feet, even though the strands are so thin we can't actually see them.

So You Want to See Your DNA

Even though DNA is too thin to be visible to the naked eye, if you had many strands of DNA tangled together—like a bowl of spaghetti—they wouldn't behave exactly like a liquid but would look more like a stringy goo. You can actually make this DNA spaghetti in your own kitchen, starting with just about any source of cells: strawberries are one popular choice, although my science lab in college had us do this experiment with liver. You can find instructions at http://learn.genetics.utah.edu/content/labs/extraction/howto/, or just search for "DNA extraction at home."

HISTONES

How does all that DNA keep from tangling up when it's in our cells? Part of the answer is that it's wrapped around tiny spools.

We can't wrap an entire strand of DNA around a single spool, though. Think of a spool of sewing thread: only the outside layers of thread are accessible, and the rest of the thread is buried in layers beneath the surface. For the cell to read the "recipes" in DNA, the molecule has to be more accessible than that.

Instead, the spools are so small that the strand of DNA can only wrap around 1.65 times. That's less than two full loops. Each of our cells has hundreds of millions of these spools, which are called **histones**. Histones, like most of the interesting things we'll meet in cells, are made of protein.

DNA or Protein?

In the first half of the twentieth century, biochemists had figured out that cells had some kind of "genetic material" that could carry information from parents to children. They even knew that it was in the nucleus of each cell. But since the material they found was a mix of DNA and histone proteins, they argued for decades over whether the DNA or the protein was the genetic material. (Spoiler: it's the DNA.)

PACKING UP

Even when DNA is wound around histones, it's still pretty big. This mix of DNA and protein, called **chromatin**, gets tightly wrapped into a cylindrical shape—still very long, but think of it more like a rope than a strand of thread. Normally, some of our DNA is in this condensed form, while other stretches of it are looser.

When it's time for a cell to divide, the ropes of DNA coil up even tighter, forming a fat sausage shape. As they scrunch up, they become visible through a microscope, and we call them chromosomes. If you've seen images of chromosomes that look like sausage links or like a big puffy letter "X", you've seen them at their most condensed.

The prefix *chroma-* comes from a Greek word meaning "color." When scientists first looked through microscopes at cells, they found that they could see everything better if they added a dye to their samples and then rinsed it off. Some parts of the cell would hold onto the dye, while others would stay clear. Scientists still do this today, often using different dyes to color different parts of the cell.

Chromatin takes up dyes easily, so when early cell biologists described what they saw, they described the sausage-like condensed DNA as "chromosomes," or literally, "colored bodies."

TRANSCRIPTION AND RNA

Copying the Recipe

We've been talking about DNA as a collection of recipes. In this section, we get to see how those recipes actually get made.

Many of our recipes, or **genes**, are instructions for making specific proteins. We'll learn more about proteins in the "What Proteins Do" section, but for now you just need to know that we make a ton of different ones. Your muscles, for example, owe their strength to proteins called actin and myosin that can contract to shorten the muscle. Your blood can carry oxygen to keep you alive thanks to a protein called hemoglobin in your red blood cells. And your digestive tract can break down food thanks to a variety of digestive enzymes produced by the stomach, pancreas, and small intestine. Each of these proteins, and thousands more, needs to be carefully built from the instructions contained in your genes.

The two major steps in making the protein are:

1. Copying the gene by creating a new molecule of RNA with a matching sequence of bases. This is called **transcription**.
2. Building the protein according to the instructions in the RNA. This step is called **translation**.

If it's hard to remember which one is which, remember that when we transcribe something in real life, we are usually writing down something that happened in a recording. Both the speech and the writing are in the same language. Similarly, the RNA is in the same "language" of nucleotides as DNA—just a different dialect, as we'll see. But converting the language of nucleotides into the language of protein is a tougher job, and it really is more like translation.

Or you can just remember that in the alphabet, the *c* (in *transcription*) comes before the *l* (in *translation*). That works too.

The Central Dogma

In 1957, Francis Crick, already famous for his role in discovering the structure of DNA, gave a lecture in which he explained a hypothesis he had: that "the main function of the genetic material is to control…the synthesis of proteins." He argued that the information in a nucleic acid like RNA can be used to make proteins, but not vice versa, and called this idea the "Central Dogma" of molecular biology. A variation on this, the rule of thumb that DNA is transcribed into RNA, and that RNA is translated into protein, is part of every genetics student's education today.

The only problem is that it's not really a dogma. *Dogma* is a religious term for a set of principles that some authority declares to be true and everybody must believe it. That's not how science works! An idea only sticks around for as long as it is backed up by evidence. Crick later admitted that he misunderstood the word—but it was too late. The name stuck.

WHAT IS RNA?

RNA is ribonucleic acid. Compare that to DNA's full name, deoxyribonucleic acid. These two molecules are very close relatives.

Like DNA, RNA is made of nucleotides that in turn are each made of a phosphate, a sugar, and a nitrogenous base. In DNA, the sugar portion is a type of sugar called deoxyribose. In RNA, it's ribose. The difference, as you might guess from the name, is that ribose has an extra oxygen atom that deoxyribose does not.

This chemical difference has consequences. RNA breaks down more easily, while DNA is more stable. (Be glad that your genome is stored on DNA!) RNA is also more flexible than stiff and stuffy DNA, so you'll find RNA in all kinds of shapes, not just a long double helix. Thanks to this folding, RNA can also base-pair with itself, so it doesn't need another strand to act as its partner. In some of the sections to come, we'll see some of the ways RNA takes advantage of these features.

Finally, RNA and DNA also differ in their nucleotide code. You'll recall that in DNA, adenine matches up with thymine, and guanine matches up with cytosine. But RNA doesn't use thymine, so adenine binds with uracil instead. (Guanine and cytosine still work the usual way.)

DIFFERENCES BETWEEN DNA AND RNA	
DNA	**RNA**
Sugar: deoxyribose	Sugar: ribose
Base pairing: A/T, G/C	Base pairing: A/U, G/C
Double stranded	Usually single stranded, but may fold back on itself in a variety of shapes
Very stable	Easily broken down

WHY TRANSCRIBE?

In our cells, most of our DNA lives in a compartment called the nucleus. But the proteins are made outside of that nucleus, in the main part of the cell called the **cytoplasm**. The process for making proteins doesn't really work with double-stranded DNA, anyway, so we copy the gene's information into a strand of RNA called messenger RNA, or **mRNA**.

Think again of our DNA as a valuable set of cookbooks. They're heavy and kind of a pain to deal with, and you certainly wouldn't want to splash spaghetti sauce on them in the kitchen. So, you leave them on the shelf in the living room and copy the recipe you want onto a scrap of paper. Time to get cooking!

Bacteria don't keep their DNA in a nucleus, but they still transcribe their genes into messenger RNA. Controlling when genes get transcribed is a good way to control when a gene's product is made, so the transcription step turns out to be handy for every form of life, whether they have a nucleus or not.

HOW IT HAPPENS

Our cells have **enzymes** whose job is to transcribe DNA into RNA copies. These enzymes are made of protein, and their shapes allow them to grab onto DNA and RNA and do all the little microscopic actions required to make the transcript.

And yes, these proteins were themselves made by the process of transcription and translation.

One important enzyme is called an **RNA polymerase**, because it can chain nucleotides together to make the polymer we know as a full-blown RNA strand. It waits for the nucleotide to float into the right spot, pairing to the DNA; then RNA polymerase locks it to the growing RNA chain so it can't get away.

There are about one hundred other proteins that work in a team to get transcription started, including **helicase**, which unzips the double helix to let the other proteins get to the DNA so they can do their work. A **complex** of RNA polymerase and some of these other proteins actually moves as it works, chugging along the DNA as it creates the RNA chain. After they finish, the proteins leave the DNA strand, and the double helix zips back up.

FINISHING TOUCHES

After we've created the transcript, we're still not done. That transcript needs a bunch of finishing touches, so other proteins and special RNAs gather around to give it a little RNA makeover.

One important step is **splicing**. Bacteria are lucky; their genes contain just the right amount of information, so their mRNAs go straight from the transcription step to translation. But ours are far too long! They contain useful chunks of genetic instructions, and we call those chunks the **exons**. But in between those exons are long stretches of RNA that don't help code for the protein. These garbage sequences are the **introns**, and the purpose of splicing is to cut out the introns and get rid of them.

So wait, why do we have introns in the first place? Often because the gene can be read multiple ways. Cut out these introns, and you have one protein. Or cut out a different set of introns, and you'll get something else entirely. This is called **alternative splicing**.

Other kinds of processing happen too. For example, a team of enzymes adds a "tail" of hundreds of adenosine nucleotides to the mRNA. This tail protects the mRNA from being degraded, or broken down, too quickly. It also helps other proteins to recognize the mRNA as something that's ready to be carried out of the nucleus.

Once the mRNA has been created, processed, and edited, it's time for the mRNA to leave the nucleus and head out to where the ribosomes are waiting.

WHAT PROTEINS DO

A Lot, Actually

An easier question might be what proteins don't do. Nearly everything in our body is made of protein, DNA, carbohydrates, or fats. The fats are to store calories, and the carbohydrates are for quick access to calories plus a few other special jobs. We already know what DNA does. So almost everything else—the structure of our body, the functions it has—we owe to proteins.

SOME PROTEINS YOU MAY KNOW

Our muscles are made mainly of protein. (That's why you have to make sure you're eating plenty of protein if you're working out and trying to build muscle.) Our muscle cells are long and skinny, and they're filled with long, thin structures called myofibrils that are made of protein. Each myofibril has two different kinds of protein, called **actin** and **myosin**. Both are made of smaller proteins glued into long ropes, but the myosin rope has little lumps on it—think of them like hands or hooks—that can grab onto the actin. When we want to contract our muscles, the myosin grabs onto the actin and pulls, over and over, like you would pull on a rope in a game of tug-of-war. Through this, the actin and myosin slide together. When we relax or extend our muscles, the myosin and actin slide back to their original positions.

The visible parts of our hair and skin are made of a tough, waterproof protein called **keratin**. Our top few layers of skin cells aren't alive; they are dead and keratinized, meaning they are filled with keratin to form a barrier between the outside world and our tender tissues. Hair is also made largely of keratin.

Proteins aren't always tough, though. **Elastin** is another protein found in our skin, and this one is stretchy, as the name implies. **Collagen** is another protein you've probably heard of, another ropelike protein that makes up a lot of our connective tissue. If your body stopped making collagen, you would start to feel the same muscle pain and loose teeth as sailors with scurvy did.

Sailors got scurvy when their diets didn't include enough vitamin C. The connection between vitamin C and collagen is, you guessed it, another protein. It's called **hydroxylase**, and it's an enzyme that does the important job of linking collagen chains together. But it needs a special ingredient, vitamin C, to be able to do its job. In fact, many of the vitamins and minerals we need in our diet are helper molecules for our enzymes.

What's an Enzyme?

An enzyme is a protein with a job. Specifically, enzymes catalyze chemical reactions. They may take apart a specific type of molecule, or they may put a molecule together. Each enzyme has a very particular job, so one enzyme can't fill in for another. We need thousands of different specialized enzymes, and thus our DNA has thousands of recipes to make them.

Digestive enzymes are probably the best-known enzymes. When you swallow food, it sloshes around in your stomach, bathing in acid. That acid sounds like it should be able to break down food, but in reality food is tough! We need specialized tools to cut the food molecules into pieces. A lot of what we eat is protein, so there are enzymes in our stomach, called **proteases**, whose job is to cut proteins apart. (Many enzymes have names ending in -*ase*.) Once the food leaves the stomach and enters the small intestine, it will encounter yet more enzymes: besides proteases to break down proteins, there are also **lipases** to break down fats and **amylases** to chop up starches into little pieces of sugar.

Receptors are another important category of proteins. Each of our cells has to communicate with trillions of others. Often, cells communicate by looking for chemical messages floating through the bloodstream. They aren't literally looking, since cells don't have eyes. But they do have receptors, which are proteins embedded in the outside membrane of the cell. Each receptor has a specific shape, just like the lock on your front door has a shape that will only accept a certain kind of key. There are receptors for growth factors, for insulin, and more. Brain cells talk to each other by releasing neurotransmitters, and how do you think the brain cell on the receiving end knows when a neurotransmitter arrives? Yup, receptors.

PROTEINS AND AMINO ACIDS

There are twenty **amino acids** that our body can build into protein. Imagine twenty different colors of candy you can string onto a necklace. If you have a specific color pattern you're trying to make, but your candy bowl is short on one of the colors, you can't make the necklace of your dreams.

Our body is the same way with amino acids. We need a full supply of all twenty amino acids to be able to build all the proteins we need. If we're short on one amino acid, we can't make the proteins that call for it.

We get some of these amino acids by breaking down old proteins we don't need anymore. But we also get proteins from the food we eat. Any time you eat a protein-rich food, like steak or tofu, your digestive system breaks down the proteins into their building blocks, the amino acids. Those amino acids are transported to our cells, where they get a new life in a brand-new protein.

WHAT'S AN AMINO ACID, ANYWAY?

Amino acids are molecules. You can build them with your model kit, or your gumdrops, as we discussed in the section titled "Atoms and Molecules."

Every amino acid has a nitrogen atom with a few hydrogens attached; that's the amino part. On the other end of the molecule, there is a carbon with two oxygens and a hydrogen, an unmistakable acid group. In between, there is also a central carbon with four different things on it (remember how carbon makes four bonds?):

- The amino group
- The acid group
- A hydrogen
- A wild card segment called the "R" group

There are twenty different kinds of "R" groups and thus twenty different kinds of amino acids.

Amino acids all have different properties. Some are hydrophilic, meaning they like to be near water, while others are hydrophobic, and prefer to cluster inside of cell membranes where there are oily molecules but not much water. Some amino acids have a positive charge, and they are attracted to other amino acids that have a negative charge. Some amino acids are small enough to fit anywhere, while others have bulky "R" groups.

Because of these wildly differing properties, if you string a bunch of amino acids together, they'll start twisting and turning so the positives and the negatives can get together, the bigger groups have room to move, and the hydrophobic parts huddle together to hide from all the water molecules. The result is a **folded protein**, and its shape after folding is determined by the sequence of amino acids and the ways they relate to each other.

Since the function of a protein depends on its shape, and its shape depends on its sequence, the instructions for the protein's sequence determine what kind of job the protein ultimately has.

TRANSLATION AND PROTEINS

Time to Get Cooking

We've made it through transcription, so now we have an mRNA with neatly edited instructions for building a protein. Next, the mRNA has to find a **ribosome**.

Each ribosome is a gigantic, multipart complex of protein and RNA. That's right—the factory that reads RNA is itself made mostly, about 60 percent, out of RNA.

There are two main pieces to the ribosome. One is larger than the other, and they look kind of like a hamburger bun, with a smaller flat part at the bottom and a larger dome-shaped top bun. The mRNA goes between them like the meat of a sandwich. (It's a lot longer than the bun, though—imagine a mile-long hot dog.)

When the mRNA gets out into the cytoplasm, all kinds of molecules bump into it. Eventually, one of those is the small portion of the ribosome. It's shaped in just the right way so it can stick on to a certain point at the beginning of the mRNA. Once that sticks, the larger top bun can join, and now we have the machinery we need to make a protein.

But we also need some supplies. Proteins are made of amino acids, so we need a bunch of those. And we also need some way of figuring out which amino acids the recipe is calling for.

tRNAS AND CODONS

Now we just need to understand how the sequence of letters in the mRNA tells the ribosome which amino acids to string together.

Since there are only four bases in RNA (A, U, G, and C) but twenty different amino acids, it takes three bases to spell out each amino acid. Each three-base sequence is called a **codon**, because it is a segment of the "code" that we can translate from the language of RNA to the language of protein. There are sixty-four possible codons, and sixty-one of them code for a specific amino acid. The other three are **stop codons**, which tell the ribosome its job is done.

So, remember those amino acids that were just floating around the cell? The ones that are ready to be built into proteins actually have something special going on. They are each attached to a **tRNA**. (The "t" stands for "transfer," because they carry, or transfer, amino acids to the protein in progress.)

A tRNA is a piece of RNA, as you probably guessed. On one end, it's attached to an amino acid. Because RNA molecules can easily fold back on themselves, a tRNA has a twisted shape that actually looks a little like a letter "T." And on one part of that RNA strand, three bases stick out, not bound to any of the other bases on the tRNA. These three bases are free to pair with any other RNA or DNA that the tRNA comes across, and it's about to bump into our nice long strand of mRNA.

THE PROTEIN-BUILDING PROCESS

Once the hamburger buns of the ribosome are attached to the beginning of the mRNA, the first three bases are available for a tRNA to bind to.

The tRNA binds to the first codon, rather than to a random spot later along the mRNA, because the ribosome has a special binding site to help it stick. The new tRNA fits into the slot on the ribosome and matches up with the first codon of the mRNA.

At this point, the ribosome chugs down the mRNA, one three-base step at a time. Each time the ribosome advances, the next tRNA that is needed finds its binding space. The ribosome connects the new amino acid to the previous ones, and in time we have a huge, growing chain. When a tRNA has done its job and no longer is attached to its protein, it gets to float away. (It may pick up another amino acid and end up doing its job all over again.)

By the time the ribosome gets to the other end of the mRNA, an entire protein has been produced. It can then fold up and be transported to wherever it needs to go—maybe another part of the cell, or maybe it will even be packaged so that it can leave the cell.

Not Every Gene Makes a Protein

So far, we've been talking about genes as recipes for proteins, but some genes just make an RNA that doesn't turn into a protein at all. These special RNAs come in many different types, but we've already seen two: the tRNAs, and the RNA components of ribosomes.

WHY DO WE NEED THE mRNA AT ALL?

- We can make several copies of the same gene and work on them all at the same time.
- The RNA will degrade over time, and when it's gone, the protein won't be made anymore. This is a way to automatically turn off protein synthesis. (If you want to keep making the protein, you can just keep transcribing it.)

- The cell can control how much of the protein it makes by turning transcription on or off. This way, the protein-building machinery doesn't have to care about when it's time to make what protein.
- While that ribosome was doing its job, a bunch of other ribosomes were likely at work, each making their own copy of the protein from that same mRNA transcript.

TURNING GENES ON AND OFF

Harder Than Flipping a Switch

Our body needs to know whether and when to make, or **express**, each of our genes. After all, we don't fill our brain cells with muscle fibers, or our skin cells with digestive enzymes. Every gene has a time and a place it's supposed to be expressed.

Gene Expression

When a gene is turned on, scientists say it is being "expressed." So, they might do experiments to ask what genes are expressed in a skin cancer cell that aren't expressed in a regular skin cell. Or they might have a gene in mind and do experiments to see what types of cells express that gene. Gene expression studies usually look for mRNA floating around a cell, because if a gene is being transcribed into mRNA, we can be pretty sure it's being expressed.

PACKING AND UNPACKING THE CHROMOSOMES

For a gene to be transcribed (see the section titled "Transcription and RNA"), the RNA polymerase, which makes mRNA, has to be able to bind to the DNA strand at the place where the gene, or recipe, starts. That's not always an easy task, since a lot of our DNA is tightly **condensed**.

When DNA is condensed, it's wrapped around histone proteins and then coiled back on itself. Imagine wrapping up strings of holiday lights onto spools at the end of the season, then packing all the spools of lights into a box, then putting the box into the attic next to a bunch of other boxes. Next time you want to use those lights, you'll have some unpacking to do.

Our chromosomes, or DNA strands, pretty much always have some portions packed up and some portions more accessible, depending on the cell's needs. When it's time to unpack a section, special proteins called activators open up the chromatin. Later, other proteins called repressors can fold it back up.

The spools that the DNA is wound on, called histone proteins, can also control whether a gene gets transcribed or not. If the histones have a certain chemical group attached in a certain place, that makes the gene on that part of DNA more likely to be transcribed. With a different group, the gene is inhibited.

CONTROLLING TRANSCRIPTION

Once the section of DNA we need has been unpacked, we still need to get the right proteins to the gene so it can start transcribing. RNA polymerase is the most important one, and it looks for a sequence of bases at the start of the gene, called the promoter, to know where to begin.

But RNA polymerase doesn't act alone. It needs a whole team of proteins to help. Some of these proteins are called **transcription factors**. One binds to the DNA to help RNA polymerase find the right spot. Others help to unwind the DNA as transcription occurs. And in addition to that promoter sequence that's right in front of the gene we want to transcribe, there is sometimes also an **enhancer** on a different part of DNA. If that enhancer is unpacked and available, certain proteins can bind to the enhancer while they also bind to the transcription complex.

How Can a Protein Bind to DNA?

Many of the ways of turning genes on and off require proteins to bind to DNA. They typically do it the same way DNA strands bind to each other! A base pair will have two or three hydrogen bonds, which are places where a negative charge on one side attracts a positive charge on the other. DNA-binding proteins have a specially shaped area that contacts the DNA, with ten to twenty places where the protein can hydrogen-bond to specific parts of the DNA sequence.

Some of these proteins can only do their job if they have been turned on themselves. For example, one protein that helps transcribe certain genes is also a receptor for a steroid hormone. When that hormone is in the bloodstream, it can enter a cell. The receptor for that hormone happens to also bind DNA in a particular spot. This means that the hormone serves as a signal to tell a cell to start expressing certain genes!

EDITING AND SPLICING THE mRNA

Once the gene has been transcribed and we have the mRNA, there's still a lot the cell can do to control what happens next.

Remember that the mRNA gets spliced before it leaves the nucleus. This means the cell's machinery cuts out unnecessary portions, which we call introns, and stitches together the remaining pieces, which we call exons. The same gene can actually be spliced in more than one way to make multiple proteins.

For example, cells in the thyroid make a hormone called calcitonin by splicing four exons together. But cells in the hypothalamus, a gland near the brain, make a different protein out of the exact same gene. Its first three exons are the same as the first three of calcitonin, but then it skips the fourth and cuts out several other pieces that the calcitonin gene throws away. The result is a different protein, which got the boring but descriptive name of CGRP, for "calcitonin gene-related peptide."

The mRNA can also be edited in smaller ways. For example, the liver makes a protein called Apo B-100 that helps to transport cholesterol around the body. It is 4,563 amino acids long. But cells in the intestine don't need to make Apo B-100. Instead, they use a protein called APO B-48 that only requires the first half of the gene. After the gene is transcribed in the intestine, a special enzyme comes along and makes one tiny tweak: it changes a cytosine in the middle of the mRNA to uracil. That has the effect of turning one of the codons that would normally call for an amino acid into a stop codon. When the transcript is being transcribed into protein a moment later, the ribosome will find the stop codon in the middle of the instructions and quit. The result: a half-sized protein that is just perfect for the intestinal cell's needs.

MUTATIONS

The Little Differences Between Us

Your DNA is not quite the same as anybody else's. Even aside from the fact that you're a mix of your mom and dad, how did your mom and dad get to be different from each other, anyway? After all, we humans all share ancestors if you look back far enough.

The answer is through **mutation**. A mutation may sound scary, but it's really just another word for change. Odds are, you have at least one place in your DNA that doesn't match up to either of your parents—it's a new mutation, all your own, that perhaps you made when you were a single cell. Or perhaps it happened when one of your parents was making the sperm or the egg cell that would later become you.

Many of these small changes are harmless. A few can be beneficial. And, sadly, some mutations can lead to diseases or disabilities that make life harder. Some mutations can even be deadly.

SINGLE NUCLEOTIDE VARIATIONS

If you compare your DNA with that of any fellow human—even a stranger—most of it will be similar, almost identical. Just about all of us humans have the same number of chromosomes, with the same genes on them. But compare your DNA to someone else's, letter by letter, and you'll find plenty of spots where you have a gene that is just one letter (or base) different. These are called **single nucleotide variants**, or SNVs. (You might also see them referred to as **SNPs**, pronounced "snips," which are *single nucleotide polymorphisms*. The difference is that a SNP is a variation known to occur in a population, usually in at least 1 percent of people, while the term *SNV* is the more correct way to refer to a variation that is super rare or whose frequency we don't know.)

Some SNVs make a big difference in how a gene works—or whether it can be transcribed or translated at all. For example, an SNV could change one of the amino acids in a protein from one that does its job to one that doesn't work as well. An SNV could even change an amino acid codon into a stop codon, so the protein can't ever be made in full. On the other hand, plenty of SNVs occur in places where they don't do much damage.

If you could compare your two copies of DNA, the one from your mom and the one from your dad, you'd find plenty of SNVs that are different between the two. But chances are, the ones you got from your dad are ones that he got from his parents. We don't make a ton of new mutations every generation, just a handful. Most of the SNVs that are different between you and a random stranger are ones that you each inherited through your families.

If you get your DNA analyzed by a genotyping service like 23andMe, it is tested to see which versions you have of the thousands of SNVs in their database. (We'll learn more about these services later in the book.)

INSERTIONS AND DELETIONS

Mutations don't have to be a single nucleotide change. Some mutations can actually make the DNA strand a little longer or shorter.

If DNA ends up missing a chunk, we call that a **deletion**. A deletion might only be one nucleotide long, so you're just missing one letter. Or it could be huge, leaving you missing half a gene, or missing a whole gene, or even missing a chunk of DNA that would otherwise contain many genes. If a deletion is large enough, it can be fatal. Some genes are so important we can't do without them.

The opposite of this is an **insertion**, where the mutation causes us to have more DNA in a certain spot than we did before. One common source of insertions are viruses that insert their DNA into ours, like a stowaway. These are harmless if they occur in a part of DNA we weren't otherwise using (the middle of an intron, perhaps). But if new DNA gets inserted in the middle of a useful gene or in one of the special areas of DNA that helps get transcription started, it could break that gene and render it useless.

FRAMESHIFT MUTATIONS

Since codons are three bases long and come right after one another, an insertion or deletion of just one base can cause a drastic change: a **frameshift**, where the start of each codon is now in the wrong place. Imagine a protein with a sequence of methionine codons: AUG AUG AUG AUG AUG. Delete one letter—the second U—and now you have: AUG AGA UGA UGA UGA UG. That would give you one methionine (AUG), one arginine (AGA), and then the very next codon (UGA) is one of the stop codons that signals the ribosome to quit. That's a drastic result from just a one-letter change.

TRANSLOCATIONS

A translocation is like an insertion and a deletion at the same time: a piece of DNA is removed from its usual place and turns up somewhere else. You still have all the same genes, and if you're lucky, the deletion and insertion didn't break up anything important. On the other hand, if you pass down the chromosome with the deleted part to your children but don't pass on the inserted part, they could be missing something important. Fortunately, translocations are rare.

MUTATIONS CAN CHANGE GENE REGULATION

Mutations don't always mess with the protein-coding parts of genes. They sometimes interfere with the things we learned about in the last section: the many clever ways we have of regulating which genes are turned on and which proteins or RNAs get expressed.

For example, remember how the RNA polymerase looks for a certain sequence of bases, called a promoter, at the beginning of a gene? Part of that promoter is known as the TATA box, because in bacteria it often has the sequence TAATTA. There are a few variations on this sequence that all still work, but some do better than others at attracting the RNA polymerase and helping it stick.

If you have a mutation in the TATA box at the start of a certain gene, the RNA polymerase will be less likely to bind. Maybe it will find the DNA but not hang on long enough to get its whole transcription factor complex together. Odds are, if the mutation is minor, the RNA polymerase will bind some of the time but just not as often. That means the associated gene will still get transcribed, but you won't have as much of it.

Just as there can be mutations in the TATA box, there can be mutations in other parts of the promoter, or in those faraway enhancer sequences. You could even have mutations in the RNA polymerase or the ribosome itself. Or perhaps there is a mutation in the interfering RNA that would otherwise serve as an off switch for a certain gene. That means the mutation would cause more of the protein (or other gene product) to be produced. Mutations like these may not change the sequence of the affected gene, but they can end up affecting how often it is transcribed.

WHAT RNA CAN DO

Lots of Nifty Jobs

DNA gets a lot of the spotlight, but its scrappy cousin RNA can do some amazing and underappreciated things.

We often think of DNA as merely storing information, while protein gets to do all the fun jobs. (Protein enzymes can glue molecules together or tear them apart, for example.) Well, it turns out that RNA has the best of both worlds: there are RNA enzymes *and* RNA information carriers.

An RNA World

There is even a theory that when life on this Earth was young and new, the *only* large biological molecules were RNA. If this is true, DNA and protein were both latecomers, and they only stuck around because they make pretty good specialists at their jobs. But in the beginning, RNA did it all.

Following are some of the jobs RNA can do, besides carrying messages as mRNA.

TRANSLATING PROTEINS

We already know that the instructions for a protein are transcribed into an mRNA (messenger RNA), and from there a ribosome translates the instructions into a protein. But what is that ribosome made of?

If you guessed protein, you're only half right. Ribosomes are made of a mixture of protein and RNA. Each human ribosome has about 6,880 nucleotides of RNA (called **rRNA**, for ribosomal RNA) and seventy-nine different proteins. (Each protein is made of many amino acids, so in total the ribosome is about 60 percent RNA.)

Meanwhile, transfer RNA (**tRNA**) molecules are what bring amino acids to the growing protein. They are responsible for actually translating the genetic code of A's, U's, G's, and C's into protein. They can do this because they are made of RNA itself, so they have three bases that match the three on the appropriate codon.

SPLICING (IN EUKARYOTES)

After a gene is transcribed, it doesn't immediately qualify as an mRNA. It needs to be processed first, and a big part of this job is done by small nuclear (residing in the nucleus of the cell) RNAs. These snRNAs (small nuclear RNAs) join up with proteins to create small nuclear ribonucleoproteins: **snRNPs**, pronounced "snurps."

These RNA/protein hybrids join up with the pre-mRNA transcript to form a complex called the spliceosome. The RNA part of the snRNP recognizes the sequences at the beginning and end of each intron, so that's a job that a protein would be unable to do. Together, the components of the snRNP identify where to cut, and then they rejoin, or "splice," the remaining sections.

RNA INTERFERENCE (RNAi)

RNA interference is a way of silencing genes even while they're being expressed. This probably started as a way of protecting against viruses.

There is an enzyme in most eukaryotes called Dicer. Its job is to find double-stranded RNA and chop it into tiny pieces. To silence a gene, then, all you need is a single strand of RNA that complements the mRNA you want to silence. This type of RNA may be called **siRNA**, for small interfering RNA; or there is another type called a micro RNA, or **miRNA**. Either way, it binds to the target mRNA, creating a double-stranded molecule for Dicer to chop up.

But we're not done. If a gene is being expressed, there won't be just one mRNA hanging around; there will be thousands. A specialized protein grabs one of those small pieces and pries apart the two strands. That single-stranded bit of RNA, plus the protein it's attached to, is called the RISC: the RNA-induced silencing complex.

The protein hangs on while the RNA component finds a new mRNA to bind to. And if the RISC is attached to an mRNA, it's not going to get translated anytime soon. Silencing complete.

VIRAL GENOMES

RNA doesn't always have to be a team player. Some viruses are so committed to RNA that they don't carry DNA in their genome at all. Measles, influenza, and polio are all examples of RNA viruses.

Some RNA viruses encode a protein called a **reverse transcriptase**, which can read RNA and produce DNA. Since that's the opposite of what our cells normally do (we transcribe DNA and make RNA copies), these viruses are called **retroviruses**. HIV, the human immunodeficiency virus, is one of the best-known retroviruses.

Once the virus has used the reverse transcriptase to copy its own genome into DNA, often its next step is to insert that DNA copy into the host's genome. Then you'll have that virus with you forever.

CHROMOSOMES AND CELLS

How We Move DNA Around

In the next few sections, we'll look at what happens to DNA when our cells divide and grow and how the DNA from two cells can come together to make the beginnings of a baby.

Each chromosome is one very, very long molecule of DNA. Since we have three billion base pairs divided into twenty-three chromosomes, each of those chromosomes is, on average, about an inch and a half long. That's a *huge* molecule.

That strand of DNA wraps around proteins called histones, which you learned about in the section titled "How All That DNA Fits Into Cells." This DNA/histone combination can then coil up, and the coiled fiber can then fold back on itself some more. Most of the time, your DNA is a mix of coiled and uncoiled segments. But when it's time for a cell to divide, every part of the DNA packs up as tightly as possible. The resulting chromosomes are shorter—they're microscopic instead of two inches long—and they're thick enough that they show up easily under the microscope.

We'll use the word **chromosome** to mean a DNA molecule that's packed up into its most condensed form. (Scientists sometimes disagree about whether the unpacked version counts as a chromosome or not. We don't need to get into that argument.) Each packed-up human chromosome has a few features:

- **Telomeres.** The DNA at both ends of a chromosome doesn't encode any proteins. Instead, it spells out a repeating sequence, like TTAGGG, over and over again. This is necessary because when DNA is copied, the copy doesn't always include the very ends of the chromosome. These repeats are there as a buffer, so we don't need to worry about losing something important. A group of specialized proteins and RNAs surround the telomeres, protecting them further.

- **Centromere.** The center of the chromosome has its own special sequence, the centromere. In humans, the repeated part spells GGAAT. The centromere can be smack in the center of the chromosome, as the name implies, but in some chromosomes, it is closer to one end than to the other. The centromere has an important job in cell division, which we'll learn about later.

- **Two arms.** The parts of the chromosome above and below the centromere are called the arms. Since chromosomes don't have an "up" and a "down," scientists have picked a way to be sure we're always talking about chromosomes the same way. We look at the chromosome with the shorter arm on top and the longer arm below. We call the top, shorter arm "p"—yes, just the letter "P." The longer arm is called "q." These two letters are always written in lower case.

- **A banding pattern.** Remember how scientists used to dye cells and see the DNA take up the dye? It turns out that not all of the DNA takes up the dye equally well. Some parts of the chromosome absorb more of the dye than others. This means that it looks stripy through a microscope, and the stripes indicate where the DNA is packed tightly together (the darker stripes) and where it is packed more loosely (the light stripes). This pattern is consistent from cell to cell, so scientists can even name specific parts on the DNA based on which chromosome, which arm (p or q), and which of the dyed bands it's in—for example, a gene at 7q3 would be on the seventh chromosome, long arm, third band from the center.

HOW MANY CHROMOSOMES?

We humans have twenty-three chromosomes from each parent, for a total of forty-six. One of the most straightforward ways to see them is to break open cells when they have the perfect packed-up chromosomes visible (in the middle of the process of dividing). When this technique, called a **karyotype**, was invented in the 1950s, the scientists would then take a photograph and cut out the chromosomes with scissors. How do you tell all those chromosomes apart? Easy. You just line them up by size.

The biggest chromosome is chromosome 1, and the next biggest is chromosome 2, all the way down to chromosome 22. Some of the chromosomes look like they're the same length, and in fact they aren't all perfectly in order. But the chromosomes got their numbers in those days, and the numbers have stuck.

What Has the Most Chromosomes?

The Jack Jumper ant only has one pair of chromosomes, and that's just females; males of this species are haploid and have just one lonely chromosome. Among other creatures, fruit flies have four pairs, and tomatoes have twelve. Starfish have eighteen pairs. Dolphins have twenty-two pairs, almost the same as us. Horses have thirty-two pairs, chickens have thirty-nine, and sharks have forty-one pairs. The black mulberry has us all beat with 154 pairs of chromosomes. But first prize goes to a single-celled protozoan called *Oxytricha trifallax*, which has 16,000 very, very small chromosomes.

Those numbered chromosomes are the **autosomes**, and most of us have twenty-two pairs of them. (Some genetic conditions, like

Down syndrome, result in a different number of chromosomes.) The twenty-third pair is either two X chromosomes (most women), or an X and a Y (most men). This pair is called the **sex chromosomes**.

Bacteria, those single-celled creatures that sometimes make us sick, do things a little bit differently. They have one chromosome, and instead of having arms or telomeres, it's just a circular loop of DNA. Bacteria also sometimes have other tiny loops called plasmids that they can pass to other bacteria they meet.

MITOCHONDRIA HAVE CHROMOSOMES TOO

Remember mitochondria? Those calorie-burning organelles inside our cells have their own DNA, and that means they each have their own chromosome. (Yep, just one.) In the sections to come, we'll learn a lot about how the twenty-three pairs of chromosomes in the nucleus divide and combine. But the mitochondria don't do anything that complex. They have a single, circular chromosome, just like bacteria do, and they divide any time they like, not just when our whole cell is dividing. In fact, a single human cell has a population of dozens or hundreds of mitochondria.

BACTERIA AND THE MICROBIOME

Germs Have DNA Too

Bacteria include some of the smallest and simplest forms of life on our planet. Each individual creature, called a bacterium, only consists of one cell. And bacterial cells are much, much smaller than typical animal cells. Remember that we have hundreds of mitochondria inside most of our cells; bacteria are about the same size as individual mitochondria, ranging from half a micron or less to ten microns. (A micron is what you get if you divide a meter stick, which is roughly three feet long, into one million pieces. For comparison, a human hair is about fifty microns wide.)

Ancient History

As far as we can tell, bacteria are the oldest forms of life that are still alive today. Somewhere along the line, a branch of their family tree became different enough from bacteria that we now put them in another group entirely. They're called archaea. From the archaea group, another set of creatures branched off. These are **eukaryotes**.

Bacteria and archaea are called **prokaryotes**, because they have no nucleus. The opposite designation is eukaryotes, which include humans as well as all animals, plants, and fungi. The word *eukaryote* means "true nucleus," because eukaryotes all divide their cells into compartments, including a compartment called the nucleus that houses DNA. Meanwhile, bacteria and archaea are tiny cells without any compartments, and they have no nucleus.

SIMPLER CHROMOSOMES

Bacterial DNA has the same structure as ours: a double helix containing the nucleotides adenine, thymine, cytosine, and guanine. But in significant ways, it's different from our DNA:

- Where eukaryote DNA is wrapped around histone proteins, prokaryote DNA is not. (Some archaea also have histones.)
- Bacteria have a lot less DNA than we do. We have three billion base pairs and around twenty thousand genes. But *E. coli*, a bacterium that lives in the human gut and is one of the best-studied bacterial species, has just five million base pairs and roughly five thousand genes.
- Bacterial DNA takes the shape of a circle. (Human chromosomes, by contrast, are noncircular lengths of DNA. They have two ends.) There are a few bacteria that have more than one chromosome, though, and bacteria can also carry smaller pieces of DNA called plasmids.
- Most bacteria only have one chromosome. They can't have paired chromosomes like us, because they don't have two parents. Each bacterial cell has only one parent.
- New bacteria are born when one cell divides. The DNA in its one and only chromosome replicates, and the two new chromosomes attach themselves to the cell membrane, at opposite ends of the cell. Then, the cell splits apart, and each is left with one chromosome. The two new cells are genetically identical to each other and to their parent.

PLASMIDS

In addition to the main chromosome, bacteria can also have tiny circles of DNA called **plasmids**. These divide so that when the cell divides, both daughter cells get their own copies of the plasmid. But plasmids can also do some other pretty cool things.

Some plasmids carry a "fertility factor," which allows a bacterium to engage in something that looks like sex. Bacteria don't reproduce sexually, but they *can* give another bacterium a plasmid. So that's a way they can transmit DNA to another individual even though their actual reproduction is asexual.

The fertility factor is a gene carried on a plasmid, and it gives a bacterium the ability to make a tiny tube called a pilus. The bacterium can then use this pilus to **conjugate** with another bacterium, and deliver a plasmid through the tube. If it delivers the plasmid with the fertility factor, the recipient gains the ability to build a pilus and transfer plasmids to other bacteria that it encounters.

If the plasmid contains other genes that benefit the bacterium, then bacteria with that plasmid will become more common in their environment. Antibiotic resistance genes are sometimes found on plasmids. That means that bacteria that are invulnerable to certain antibiotics can pass on that trait to other bacteria they meet—including, potentially, pathogenic (dangerous) germs.

So, plasmids can be bad news. But they can be good too! Since a bacterium will replicate whatever plasmids live in it, scientists can create a plasmid with a gene that they are interested in studying. They give that plasmid to some bacteria, and from then on, whenever the bacteria replicate, they are replicating the plasmid too. Since the bacterium is carrying the plasmid with that gene, it's probably also expressing the gene on that plasmid. If you want to make a lot of some protein, you can put the gene for it into a plasmid and transfect a bunch of bacteria with it.

This is how insulin is made for people with diabetes. Insulin is a protein that acts as a hormone. Scientists have copied the human gene for insulin, stitched it into a plasmid, and given that plasmid to a bunch of bacteria. Those bacteria, grown in huge tanks in a factory, end up producing huge amounts of insulin. That's where diabetic people get the insulin that they need to regulate their blood sugar.

THE MICROBIOME

Bacteria live everywhere in the world (yes, even Antarctica). They teem in every handful of dirt, every gallon of ocean water. Not to gross you out, but every surface you can see or touch—including this book—is covered in bacteria. Your skin is coated in an invisible layer of the little guys. Your mouth is full of them, and your digestive tract contains about as many bacterial cells as there are cells in the rest of your body. (The bacterial cells are smaller; that's how we can accommodate so many.)

The population of bacteria living in any one place is known as the **microbiome** of that place: for example, the bacteria and other microscopic creatures that live in our digestive tract are known as the gut microbiome.

Since there are so many different bacteria, healthy microbiomes usually include many different species that all interact with each other in an ecosystem. Imagine a rainforest, where trees make leaves, bugs eat the leaves, frogs eat the bugs, and so on. Rainforests have a rich diversity of species doing all kinds of different jobs and relating to each other in different ways. Maybe some of the animals fight with each other; maybe sometimes there is a symbiosis with two species helping each other out.

Microbial ecosystems are like this too. One species of bacteria might make a chemical as waste that another likes to eat as food, for example. If we get a diarrheal disease, our microbiome might be thrown out of whack, with some species taking over and others dying off. Or if we take a probiotic pill, the "good bacteria" from the pill might interact with our permanent residents in a way that helps them shape up and take better care of the person they live inside. (Bacteria in probiotics don't usually stick around for very long; that's why they come with more than one pill to a bottle. You have to keep taking them if you want the benefits they give.)

STUDYING BACTERIA THROUGH THEIR DNA

Now that we have plenty of techniques to sequence DNA, we understand more about the microbiome than ever before. Microbiologists used to have a tough job ahead of them if they wanted to figure out what species of bacteria lived in a certain place. They would have to gather samples, perhaps rubbing a cotton swab over a fecal sample if they were studying gut bacteria. Then they would have to rub that swab onto a Petri plate or swish it in a flask of bacterial growth medium. Bacteria would start to grow on the plate or in the flask, and the scientists would have to do careful work to examine what kind of bacteria they seemed to be.

However, now that DNA sequencing is (relatively) cheap and easy, scientists take that swab and analyze its DNA. Every bacterium has ribosomes, just like we do, and one portion of that ribosome is called the 16S rRNA. It turns out that all bacteria have a 16S gene, but the exact sequence of this gene varies from one type of bacteria to another. Scientists can sequence this one gene for an entire swab's worth of bacteria at once.

This technique is not just quicker than the old way; it also gives more thorough results. If you're going to grow bacteria on a Petri plate, they would have to be bacteria that thrive in the presence of oxygen. But plenty of bacteria die if they contact oxygen. (Remember, there's not a flow of air through our digestive tracts; we breathe with our lungs, not our butts.) With sequencing, we don't need to know the special care and feeding requirements of each type of bacteria; we can just check out the sequence and know what species are there.

Where the Microbiome Is

We have bacteria all over our bodies, but some body parts house more than others. Our stomach, thanks to its acidic environment, only has about ten million bacteria. Our skin, in total, supports one hundred billion bacteria. And our large intestine, the main residence of our gut bacteria, contains a whopping one hundred *trillion* bacteria.

VIRUSES

Little Packages of DNA or RNA

Viruses are even smaller than bacteria. In fact, scientists don't always agree on whether they should count as forms of life or not. For now, it's probably most accurate to say they are not truly alive. However, they have their own genetic material—sometimes DNA, sometimes RNA—so we'll at least give them their own section.

Viruses only have two components: a protein shell and a nucleic acid payload. The viruses that cause smallpox, herpes, and chicken pox all contain double-stranded DNA as their genetic material. In that way, they're like us. Others have RNA as their genetic material. HIV is one example of an RNA virus.

Whether a virus's nucleic acid is RNA or DNA, it's enclosed in a package made of protein. That's all a virus is: just packaged-up nucleic acid. You could argue that it's not a living creature, just a packet of chemicals.

However, when a virus infects a cell, those proteins and nucleic acids spring into action. They hijack the cell's own machinery, so that instead of transcribing and translating its own DNA, the cell uses its precious resources to transcribe and translate the virus's genes.

SOME VIRUSES AND THEIR TYPICAL GENOME SIZES			
Influenza	single stranded	RNA	14,000 bases
Measles	single stranded	RNA	15,894 bases
Adenovirus (common cold)	double stranded	DNA	36,000 base pairs
Rotavirus (diarrheal disease)	double stranded	RNA	18,000 base pairs
Variola (smallpox)	double stranded	DNA	186,000 base pairs

HOW A VIRUS INFECTS A CELL

First, the virus approaches the cell.

Many viruses take the shape of an icosahedron, so they look like little jagged circles under an electron microscope. (At just a few nanometers wide, most are too small to be seen through a regular light microscope.) Every virus has a preferred cell, or type of cell, that it likes to infect. Some viruses that infect bacteria, called **bacteriophages**, have a slightly different structure. On the bottom of the icosahedron-shaped package, they have a tube and six legs that appear to stick out like a spider's legs. The whole thing has the appearance of the spacecraft that landed on the moon in 1969.

The virus has to find a way to attach to the cell. Bacteriophages can inject their DNA directly into the bacteria they infect, but many of the viruses that infect humans have proteins on their surface that recognize specific proteins that they expect to encounter on the surface of human cells. They attach to those proteins and try to trigger them to allow the virus in. For example, a rhinovirus, one of several viruses that can cause the common cold, tricks the cell into thinking it's time to ingest something and bring it inside the cell. The virus then tries to break out of its little digestive chamber. Most cells aren't fooled, or if they do get infected, our immune system manages to kill off the infected cell before the virus can spread. But if we get sick, that's because some virus managed to get through.

HOW A VIRUS REPLICATES

Once a virus's DNA or RNA is inside the cell, it takes advantage of the cell's tools for transcribing and translating genes.

If the viral genome is made of DNA, and it's infecting a eukaryote like a human, that DNA has to reach the nucleus. That's where our transcription machinery is, after all.

If the virus is a retrovirus that contains RNA, it encodes a reverse transcriptase. Ribosomes find the viral RNA and translate it, creating the reverse transcriptase that can then make a DNA copy of the viral genome.

Either way, a successful infection results in the virus's genes being transcribed and translated in much the same way as our own genes. In one version of the viral life cycle, called the **lytic cycle**, the cell is soon using all its resources to make new viruses, and the viruses eventually burst out of the cell, killing it.

But there is another way: the **lysogenic cycle**. In this case, the virus inserts its DNA into the host cell's genome. Then it may lie in wait, getting copied whenever the host cell's genome gets copied. One day in the future, it can activate and cause a new batch of viruses to be made.

This is what the virus that causes cold sores does. You can't get rid of the virus, but you'll only notice outbreaks when the virus decides to take action. This happens most often when the host immune system lets its guard down—for example, when you're stressed.

Just like a squirrel might bury a nut in the fall and forget to come back for it, some viruses that went the lysogenic route "forgot" to ever pop out and make us sick. Over the millennia, we've actually acquired quite a few defunct viruses in our genome, comprising somewhere between 1 percent and 8 percent of human DNA.

MUSHROOMS AND YEAST

A Strange Kingdom

Fungi are one of the less appreciated kingdoms of life, but they are weird and wonderful. You've certainly seen them in your life and probably have some nearby right now.

If you're reading this at lunch and you're having a portobello mushroom sandwich, or a burger with mushrooms and onions, those mushrooms are fungi. If you're outdoors, look down: the soil is full of microscopic fungi and perhaps also some wild mushroom species. If you're indoors, you'll see them too: the mold on bread and the mildew in your shower are also in kingdom Fungi. To bring this to an even more personal level, do you remember your microbiome? The ecosystem of microscopic creatures that live in you and on you and especially in your intestines? Tiny fungi are part of that ecosystem.

Fungi eat food. They may look like plants sometimes, but they can't live off sunlight and air like plants do. Fungi secrete chemicals that dissolve whatever is around them—a rotting log, perhaps, or a loaf of bread—and then they absorb the nutrients that are released. If you leave some food in the refrigerator too long and find mold growing on it, that mold is doing the same thing you do: eating that delicious food.

YOUR FRIENDLY NEIGHBORHOOD MUSHROOMS

When you think of a fungus, you probably think of mushrooms. But a mushroom is actually just the reproductive organ—technically, the "fruiting body"—of a larger organism. If you see a mushroom growing out of the ground, then underground you know there must be a network of hyphae. Hyphae are thin filaments that branch and form a mass called a mycelium. The hyphae are made of fungal cells. The fruiting body is also made of hyphae, tightly packed together into a solid structure.

Most fungal cells have haploid nuclei, meaning they have just one copy of each chromosome. When hyphae from two different individuals meet, their cells can merge, but their nuclei do not. These dikaryotic ("two nuclei") cells keep on dividing and eating as normal, until it's time to have babies. Then they work together to make a fruiting body.

Sex Is Optional for Mushrooms

The fungi that make mushrooms don't *have* to make mushrooms. They can just keep growing, making more and more cells asexually (whether they are haploid or dikaryotic). And when food is plentiful, that's often what they do.

However, when food is scarce or things are looking desperate, the fungus is more likely to try making a fruiting body and dispersing some spores. Since these spores carry a mix of their two parents' DNA, they are a little different from either of their parents. This is how fungi mix and match their DNA for greater variety.

The fruiting body is made of dikaryotic hyphae, except for the spores, which are found in the gills of the mushroom. There, the two nuclei in each cell fuse to make a cell that is momentarily diploid. Then this cell undergoes meiosis, splitting into four haploid cells. Each of these haploid cells becomes a spore. The spores then make their own way in the world, either puffed into the air or brushed onto a creature passing by. When a haploid spore finds the right environment, it starts growing its own haploid hyphae. It can grow this way for a while, or it can meet up with another haploid mycelium.

YEAST: A MODEL ORGANISM

All fungi are eukaryotes, even the microscopic ones. That means they have compartments in their cells, including a nucleus to hold their DNA. Microscopic yeasts are some of the simplest eukaryotes that scientists can grow in the lab, so there are a *lot* of scientists studying genetics and cell biology in yeast. (Specifically, they use the species called *Saccharomyces cerevisiae*—the same species used in brewing beer.) Yeasts are a lot easier to take care of than other eukaryotes like fruit flies or mice or people. If you're studying a process that happens in all eukaryotes, you'll probably keep it simple and start with yeast.

This species of yeast also has a small genome: just twelve million base pairs. Back when sequencing was expensive, laboratory yeast was among the first few species whose genome was fully known. Yeasts have a lot of the same "housekeeping" genes that we do, meaning genes that provide the instructions for boring, basic tasks that every cell needs to do. We actually have a lot in common with yeast.

While bacteria only have one chromosome and reproduce asexually, yeasts can have sex if they want to. Yeasts are unicellular fungi. They don't make hyphae. Instead, cells live alone. These cells can be diploid or haploid, and they divide by mitosis. In a colony of yeast, most cells are diploid.

Diploid cells reproduce by mitosis as long as times are good, but when they're stressed or food is scarce, they take a gamble on making offspring. They divide by meiosis, forming four haploid cells.

Haploid cells can also live alone, but they are strongly attracted to other haploid cells. They come in two "mating types," called a and α. They sniff each other out by means of a pheromone, or chemical attractant, that one produces and the other responds to. When two cells of opposite mating types meet each other, they can fuse together to form a diploid cell.

What if you're a haploid yeast cell, but there are no other haploid cells around of the opposite mating type? Lucky for you, it's possible to switch types. If you're α, you can become a so you have a chance to mate with one of the a types around you. This way, the full life cycle is still possible.

All this switching and recombination can be annoying, though, if you're a researcher trying to study yeast with a particular set of genes. So, yeasts grown in the laboratory have often had their DNA altered so they can no longer switch mating types.

PLANTS AND CROPS

Never Look at Flowers the Same

When you yank a dandelion out of a crack in a sidewalk, it may be hard to imagine that you have much in common with that weed. But you do—on a genetic level, at least. Plants have chromosomes just as we do; most of the time, they have two copies of each, just like us. They also combine those genomes much as we do, mating with other individuals to pass their genes to their offspring.

You're probably most familiar with plants that produce seeds, which make up 90 percent of plant species alive today. We'll get to those in a minute, but first let's take a look at some other ways that plants merge and spread their DNA.

MOSSES

Mosses are a group of plants that never developed the vascular tissues that other plants use to carry water from roots to tip. That makes mosses soft, leafy, and short. This stage of the moss's life is called a gametophyte, and its soft spongy parts are actually tiny leaflike structures made of haploid cells. That means that when you look at a carpet of moss in a forest, you're looking at creatures that only have one of each chromosome.

The haploid cells of this leafy plant divide by mitosis to make gametes: either egg cells that are nestled in a pocket in the plant, or tiny swimming sperm. Yep, some plants have sperm. These plants tend to live in swampy areas, because the sperm have to be able to swim to find a mate.

When a sperm meets an egg, the two cells fuse to make a zygote. The zygote, still attached to the parent plant, divides by mitosis to make a structure of diploid cells. If you look very closely at a tuft of moss, occasionally it will have tiny structures sticking up like lollipops. Inside these lollipops, or sporangia, cells undergo meiosis to make haploid cells called spores. Those spores can land on the ground nearby, where they can undergo meiosis to make a whole new leafy, haploid, soft fuzzy bed of moss.

FERNS

Ferns also have haploid and diploid stages, but in this case the large, visible part of the plant is diploid. (Most plants are diploid in their most visible life stage; mosses and their relatives are the exception.)

Look at the underside of a fern's fronds during the right time of year, and you may see fuzzy brown dots called sporangia. Some ferns, like the ostrich fern, make an entirely separate, brown frond that is covered in sporangia without any green bits. To fill the sporangia with haploid spores, some of the fern's cells undergo meiosis. So far, they seem a lot like us.

But their spores don't immediately join with another plant's spores. Instead, when the spores land on a good spot, they begin growing, the cells dividing by mitosis. The result is a gametophyte made of haploid cells, and it often looks like a tiny heart-shaped leaf. This gametophyte can produce eggs and sperm, and again the sperm have to swim through a moist environment to reach the egg. When the sperm and egg meet, they fuse into a zygote, and this zygote divides by mitosis to grow into yet another large fern.

PINE TREES AND OTHER CONIFERS

Land plants don't rely on sperm swimming through swampy ground. They have **pollen** instead.

Pine trees are diploid. They make male and female pinecones, and in each cone, cells undergo meiosis to produce the gametes. Male cones make pollen grains, and female cones contain ovules. After pollen reaches the ovule, the ovule completes meiosis, and the pollen grain grows a tube leading to the egg, where it releases sperm. These trees still have a haploid generation, but it occurs very briefly inside the pinecone. The resulting seed at the base of each pinecone scale is diploid. When it lands on the ground and germinates, it undergoes mitosis to grow into a pine tree.

Pollination

Pollination isn't just a job for bees. Pine trees are wind pollinated, and they make enormous amounts of tiny pollen grains that can float on the wind. Plenty of flowering trees and grasses are wind pollinated too, including corn (maize) plants. Pollination by animals can include bees, but also (depending on the plant) butterflies, moths, ants, flies, beetles, hummingbirds, and even bats.

FLOWERING PLANTS

Flowers are sex organs. Let's take a tulip as an example: if you look inside you'll see one column standing up in the middle of the flower, with a little cloverleaf shape called the stigma at the top. That column is the female part, the carpel. Surrounding it are usually six stamens, each consisting of a thin stalk with a pollen-covered anther at the top. Those are the male parts of the flower.

Some flowers have both female and male parts, like the tulip. Some of those, like Mendel's pea plants, have no need to open up to the outside world; they fertilize themselves, transferring pollen from the anthers to the stigma as the flower develops. Other plants separate their male and female parts into different flowers. And some pick one sex and stick with it: a female Osage orange or ginkgo tree will never make male flowers; it has to trust that there will be a male tree somewhere nearby.

Like pine trees, flowering plants are typically diploid, and they make haploid cells that act as gametes. These cells combine when the flower is fertilized, and the diploid zygote develops into a seed. The ovary that surrounds the seeds develops into a fruit. Apples and oranges are fruits, but so are some vegetables, like peppers and zucchini. The burrs that stick to your clothes when you walk in the woods are also fruits. Botanically speaking, nuts and grains are also fruits. Each kernel on an ear of corn is a fruit. The white fuzzy things that float away when you blow on a dandelion puff? Also fruits.

A Model Plant

Arabidopsis thaliana is the plant that biologists find easiest to study. It's a weed in the mustard family that only grows about four to eight inches tall. It also has a small genome: just 135 million base pairs in five chromosomes.

MAKING MORE CELLS

Just a Division Problem

One cell is never enough. If you're a single-celled creature, such as a bacterium, eventually you will want to make more of yourself. If you are a multicelled creature, you'll need to grow.

You started out as a single cell, made from a merger of sperm and egg. Then your one and only cell divided, and you had two cells. Each of those divided, and you had four. Fast-forward a few dozen times and here you are, a conglomeration of trillions of cells.

You haven't finished dividing, though. Every day, cells in the deep layer of your skin divide, so that you have a never-ending supply of cells to make up the outside layer of your skin. If you get a cut, the cells will divide even faster to cover the gap. Meanwhile, your stomach acid constantly eats away at the cells of your stomach lining, so your stomach cells are constantly dividing too. If you're young enough that your arms and legs are still growing, each of those long bones has an area called a growth plate, where special cells divide to create more and more bone to make you taller.

Each time a cell divides, it has a ton of work to do. Think of all the things in a human cell: the mitochondria, the ribosomes, all the little compartments, all of the protein scaffolding that helps the cell hold its shape. (Yep, cells have skeletons too.) All of that needs to get divvied up between the two smaller cells that result. One of the trickiest things for the cell to split up is the nucleus, with all of its spaghetti-like DNA.

ONE FOR YOU, AND ONE FOR YOU...

Splitting up the DNA in a human cell is not an easy job, and it would be even harder if the DNA stayed in its usual form like a big pile of spaghetti. Fortunately, when the cell is getting ready to divide, it expresses proteins and RNAs that help package the DNA into those dense structures we learned about, the chromosomes.

There's another special thing that happens during that preparation stage. The entire cell is getting bigger, accumulating more of everything. After all, the finished cells will each be half the size of the original cell, so most cells enlarge before they divide to give their daughter cells (yep, that's the technical term) a head start.

During this stage, the DNA in the nucleus is making its own preparations. Specifically, it copies itself, or **replicates**. That way, there are two copies of the DNA, one for each of the daughter cells.

Replication

When DNA copies itself, the process is called replication. It looks a little like transcription: the DNA strand has to unzip, but now instead of an RNA polymerase coming in, it's a DNA polymerase. This enzyme knows how to make DNA that matches one of the now-exposed strands. Another copy of DNA polymerase will do the same thing on the other side of the ladder. In fact, it takes many, many DNA polymerases, all working at the same time, to copy our entire genome.

Let's take a minute to count what exactly has doubled. But first, here's what has *not*. A regular cell, not dividing, has forty-six chromosomes. (Twenty-three of each, if you're genetically female. Males still have forty-six chromosomes, but with one mismatched pair. More about that in another section.) Each of those forty-six chromosomes is made up of a strand of DNA that is in a "double helix" shape. All of that is one full set of DNA for one plain old regular human cell.

But when the DNA replicates, each of those forty-six strands gets copied. So now you have *two* copies of everything from your mom, and *two* copies of everything from your dad. That sounds like it should mean we have ninety-two chromosomes for a brief minute, but that's not exactly true.

JOINED AT THE HIP

When a chromosome's DNA replicates, it doesn't result in two separate strands. The middles of the two strands stay stuck together, like conjoined twins. The resulting chromosome is in the shape of a letter "X."

Each of the two new strands is called a **chromatid**, and the middle part, where they are joined, is the **centromere**. This whole division situation is why chromosomes have a centromere at all. When the cell is preparing to divide, the centromere attracts a group of proteins called the kinetochore. And the kinetochore is what helps the DNA get doled out to each new cell equally.

THE JOY OF DIVISION

The process that splits chromosomes is called **mitosis**. In humans and other mammals, the membrane around the nucleus, called the nuclear envelope, disintegrates. (Other creatures, if they have a nucleus, can usually split up the chromosomes inside the nucleus, and then the nucleus itself divides in half. But no, we had to make it complicated.)

With the nuclear envelope gone, structures on both sides of the cell start building long protein tubes that connect the kinetochores around the chromosome to structures called centrosomes on either side of the cell. The collection of tubes is called the mitotic spindle.

As the cell divides, the spindle is what does the work of separating the chromosomes. A microtubule from each side of the cell grabs onto each chromosome at the kinetochore. Other tubules grab onto their counterparts coming from the other side of the cell. Then, as the microtubules push on each other to push the two ends of the cell apart, the microtubules that are connected to chromosomes begin to shrink. Since they are firmly attached to the chromosomes, the chromosomes get pulled apart.

This is how we can be sure one chromatid goes to each end of the cell. Chromosome 1, for example, will have microtubules from each side of the cell attached to it. It gets torn apart, with one chromatid going to one daughter cell, and the other chromatid to the other. Now each cell will have its own chromosome 1. The same thing happens with all the other chromosomes.

Since this is just mitosis, our ordinary cell division, nothing special happens with the sex chromosomes. If you're a woman with the usual XX, you'll have two X-shaped X chromosomes, just like you have two X-shaped chromosome 1s and two X-shaped chromosome 2s, and so on. If you're a man with XY sex chromosomes, you'll have an X and a Y chromosome, each doubled up and temporarily existing in that X-shaped state.

Once each daughter cell has one of each chromosome, the microtubules of the spindle disintegrate, and the nuclear envelope re-forms around each of the two new nuclei. The rest of the cell finishes dividing, and now we have two times more cells than when we started.

That's how a cell divides—how we get two cells where we started with just one. The new cells are each roughly the same as their original parent, with an identical copy of all their parent's DNA.

DNA REPLICATION

How Copies Are Made

For cells to divide, they need to double their DNA so each of the new daughter cells can have their own copy. How exactly do they do that? Let's dive in.

Remember that the two strands of DNA are matched up in opposite directions. (If you imagine them both as people, one is standing on their feet, while the other is standing on their head.) We call the two ends of each strand the 5' ("five prime") and 3' ("three prime") ends.

One thing that makes DNA replication a little bit complicated is that the enzymes that make DNA can only work in one direction: they can only add nucleotides to the 5' end of the strand, not the 3' end. And here's another problem: the enzyme that does the main part of that job—it's called a DNA polymerase—can only work from a *double strand* of DNA. It can't just add new nucleotides in the middle of an existing single strand.

Before we talk about humans, let's take a look first at how bacteria replicate their DNA. *E. coli*, a species of bacteria that scientists love to study, has a single chromosome with about 4.2 million base pairs. Bacterial chromosomes are typically circular, so they don't really have "ends" at all. There's a single point on the chromosome called the origin of replication, so when the bacterium wants to make more of itself, a complex of proteins begins at this origin and separates the two strands of DNA there.

To get DNA synthesis started, we need a double strand. Fortunately, an enzyme called primase comes to the rescue. It can lay down a dozen nucleotides of RNA to match whatever DNA is on the single strand. The DNA polymerase can then come in and see a double strand it can work with. (That double strand is half DNA and half RNA, but that's no big deal; other proteins will swap out the RNA for DNA later on.)

You need more than one DNA polymerase to replicate all the DNA. Think about what happens if you begin pulling the two strands of DNA apart: you get a little area where they are separated, called the **replication bubble**, and at each end of this is a shape we call the **replication fork**. We need to duplicate both strands on both sides of the replication fork, so that takes a minimum of four sets of replication machinery.

But it's actually a little more complicated than that. Let's look at these enzyme complexes. One starts at the origin and speeds down the DNA as the bubble grows longer, adding nucleotide after nucleotide in the 5'-to-3' direction. A second complex can do the same thing, going down the *other* strand in the *other* direction. At each place where these complexes are working, the matching strand needs to be replicated too.

How Long Does It Take?

Eukaryotic DNA polymerases can lay down 500 to 5,000 nucleotides per minute. At that rate, replication would take a week per chromosome. Fortunately, our genome has 30,000 origins of replication, so we can replicate from a lot of different places at the same time. The process can finish in just minutes to hours.

Let's zoom in to just one of the replication forks. (Whatever is happening here is also happening at the other side.) One DNA polymerase is just synthesizing and synthesizing as the replication fork opens. On the other strand, though, the DNA has to be synthesized in the opposite direction, heading *away* from the replication fork. But as soon as it gets started, the fork has opened up further, so another primase and DNA polymerase start in on the job, again moving away from the replication fork. We call this strand of DNA the "lagging" strand, because little snippets of DNA (called **okazaki fragments**) are being produced as the polymerases try to keep up with what's going on over on the "leading" strand. Eventually, all the little gaps on the lagging strand get filled in.

STEPS IN HUMAN DNA REPLICATION

Human DNA is a little more complicated than bacterial DNA. That's partly because bacteria are so simple, so whatever they do with a few enzymes we do with a whole bunch more. But human DNA is also much larger than bacterial DNA. Here are the steps:

1. **Initiation:** The human genome has about thirty thousand places where replication can begin. When the cell is ready to begin replicating DNA, a licensing protein binds to each of these origins. Later in the cell cycle, the replication machinery initiates replication but only at licensed origins. After replication begins at an origin, its licensing factor is removed so that that origin won't be replicated again. (Otherwise you could end up with chromosomes where some parts are replicated more than once, making a branched network of DNA instead of two linear molecules.) Another protein called geminin takes the licensing factor's place. At the end of mitosis, geminin is

destroyed, allowing replication to begin again the next time the cell divides.

2. **Elongation:** As the replication fork proceeds down the chromosome, topoisomerases unwind the DNA so it doesn't get tangled up. (Bacteria do the same thing, often with a different enzyme called a gyrase.) Our DNA polymerases work a bit differently than those in bacteria; we have one polymerase that can make its own RNA/DNA hybrid primer, and then it's followed by different DNA polymerases that just make DNA. One type works on the leading strand and another on the lagging strand. We also have a special DNA polymerase whose job is to replicate mitochondrial DNA.

3. **Nucleosome assembly:** Because we have histones to organize our DNA, we have to add them in to new strands of DNA as needed.

How Plasmids Replicate

Plasmids have a simple and clever solution to replication. One of the strands serves as a template, and the other strand gets a snip in its backbone. A polymerase hops on and lays down new DNA as the now-redundant original strand peels off. This can happen as many times as necessary, making a bunch of linear copies. (It's one of these linear copies that gets injected into the other cell during bacterial conjugation.) Finally, each of these strands joins back into a circle, and DNA polymerase returns to make its matching strand.

MAKING SPERM CELLS

Half the DNA, All the Fun

Where do babies come from? Sperm and eggs, of course. But where do sperm and eggs come from? A special type of cell division, it turns out. Gametes like sperm and egg cells only have *half* of a full set of DNA. Remember how you get half of your chromosomes from your mom and half from your dad? That's because each of your parents only ever gave you half a set. But when you put your mom's and your dad's chromosomes together, you have the full set of forty-six.

HOW SPERM IS MADE

Healthy human testes make about one hundred million viable sperm every day. But the entire process of making sperm takes a long time—about seventy-four days. (Many batches of sperm are made at once.)

Inside the testes are long squiggly tubes called seminiferous tubules, and inside those are stem cells called spermatogonia that constantly undergo mitosis, the process of cell division we learned about in the last section. When one of these stem cells divides, it just makes more cells identical to itself. This process keeps a steady supply of this type of cell.

What happens next, though, is special. The spermatogonia undergo a two-step process called **meiosis** to create sperm cells. We need a special process, rather than just another round of mitosis, because sperm cells need to be **haploid**, meaning they only carry one of each chromosome pair—twenty-three individual chromosomes, not forty-six.

MEIOSIS

The first part of meiosis starts with making a full second copy of DNA, just like we did for mitosis, *even though* our goal is to have cells with half the original DNA. That means our cell will need to divide twice, making a total of four cells. Each cell will have half of the original cell's DNA.

So the DNA in each chromosome duplicates, resulting in X-shaped chromosomes where the two identical strands are joined in the middle. But instead of dividing right away, the cells spend a little bit of time in this first stage, called **prophase I**, letting the chromosomes' DNA mingle with each other.

That means that if you are the one making sperm, your chromosome 1 from Dad gets duplicated, and your chromosome 1 from Mom does too. Then Mom's and Dad's copies sit around for a little while, and while they do, they link up with each other. Dad's duplicated chromosome sticks on to Mom's. The same happens for all the other pairs. The X and Y chromosomes don't have perfect partners, so they pair up together. There are portions on the X chromosome that are similar enough to the Y that these two can stick to each other.

FINISHING CELL DIVISION

We can then proceed with the first division, which we call **meiosis I**. This time, instead of the microtubules ripping apart those X-shaped duplicated chromosomes, they pull the chromosome *pairs* apart. Mom's replicated chromosome 1 goes to one cell, and Dad's replicated chromosome 1 goes to another.

Now, you don't end up with one cell having all of Mom's DNA and the other having all of Dad's. The mix is actually random: maybe in one cell, chromosomes 1 and 2 are from Mom, chromosome 3 is from Dad, and chromosome 4 is from Mom again. Which ones end up in which cell is just the luck of the draw.

The cells finish dividing, and now you have two primary sperm cells. Each has just twenty-three chromosomes, but each of those twenty-three is an X-shaped chromosome with two identical double helices. Our work is only half done, but at this point you can tell that one of the cells is going to make sperm cells that can become a boy (since it has a Y chromosome and no X), and the other will only be able to make girls (it has an X chromosome, with no Y). We'll learn more about X and Y chromosomes in a bit.

Mitosis versus Meiosis

Don't let the names fool you: these are two different types of cell division! In mitosis, a cell starts with twenty-three pairs of chromosomes and produces two daughter cells with the same twenty-three pairs of chromosomes. In meiosis, the cell still starts with twenty-three pairs, but by the end there are *four* cells, and each has twenty-three *single* chromosomes—no pairs.

Now it's time for the second cell division. This one looks more like mitosis: each X-shaped chromosome gets microtubules stuck to it, and the microtubules rip the X-shaped chromosome apart. Since each of the two cells divided, we finish with four. Each of those four cells only has twenty-three lonely, single chromosomes. The four cells then get a spermtacular makeover, giving them long tails and a characteristic spermy shape, and then they are ready to go out into the world.

SHUFFLING DNA

Let's say you're a guy, making a whole bunch of sperm. Whatever DNA you pack into each sperm cell will be half of your future child's genome. But you aren't going to make a bunch of *identical* sperm, right? That would be boring. It would also mean that all of the children you have with one partner would have the same DNA, like identical twins.

There are a few ways to make your sperm all different from each other. One is that the chromosomes mix up randomly, so that a given sperm might have the chromosome 1 you originally got from your mom, and a different sperm might have the version of chromosome 1 that came from your dad. You know you'll always be passing down half your DNA in each sperm, but you don't know exactly which half.

But there's another way that the DNA gets shuffled. Remember how the two copies of each chromosome (say, chromosome 1) stick together during meiosis I? During that time, the two strands of DNA aren't just hanging out. They actually intertwine so closely that they can swap pieces. When they pull apart, that chromosome you initially got from your dad might have a chunk of your mom's DNA in it, and vice versa. This is called crossing over, or **recombination**.

MAKING AND FERTILIZING EGG CELLS

Your Humble Beginnings

Egg cells also do meiosis, but it looks very different. For one thing, we only release one egg cell at a time—not four, and definitely not one hundred million. If you thought seventy-four days was a long time for a cell to divide a few times, egg cells will put that in perspective: for them, meiosis takes decades.

The stem cells that will become egg cells, called oogonia, actually do their last round of mitosis (normal cell division) while the person with the ovary is developing, themselves, in utero. The oogonia have to undergo meiosis, just as sperm cells did, to produce an actual egg cell. However, the process stops as soon as the cell begins to go through meiosis. By the time a baby girl is born, the cells inside her ovaries have been frozen in the very first stage of meiosis. And they stay like that for decades.

In that newborn baby's ovary, the DNA in each of her egg cells has already duplicated, and the chromosomes are already in an X shape, and the two Xs of each pair are stuck together. That's where the process pauses. Those X-shaped chromosomes will snuggle together for years, until this little girl hits puberty.

SUSPENDED ANIMATION

The egg-cell-in-progress (called a primary **oocyte**) might wait in mitosis I, chromosomes stuck together, for decades. When the person hits puberty, one of those oocytes will mature and finish dividing. That's a wait of perhaps thirteen years. And when the person hits menopause, her very last oocyte to mature will have waited even longer—perhaps fifty years.

Remember how the chromosomes swap parts while they're snuggled up? To do this, they have to be attached very securely to each other. Occasionally, after being stuck together for so long, they don't separate cleanly.

That's why the risk of a condition like Down syndrome increases with the age of the mother. The longer the chromosomes were together, the more likely they will experience the rare case of **nondisjunction**, or not separating properly. The egg cell that results may end up with an entire extra chromosome, so that the child ends up with three copies. Down syndrome occurs when a child has three copies, or a **trisomy**, of chromosome 21.

Most other trisomies don't develop into an embryo. And if an egg cell ends up missing the chromosome completely—if nondisjunction left it with zero rather than double the usual number of a certain chromosome—that cell is also unlikely to lead to a viable pregnancy.

There are also cases where recombination occurs and the chromosomes separate just fine, but during the recombination, the DNA swap didn't occur quite perfectly. Maybe one chromosome gives a little more than it gets, or vice versa. The result can be that the child is missing a piece of a chromosome or has an extra piece. Whether this is a serious condition or not depends on what genes were affected.

WAKING UP AGAIN

From puberty until menopause, only about ten to twenty of these oocytes will wake up each month and finish the process of meiosis. Those few cells begin producing hormones. Whichever makes the most hormones the fastest is the winner and gets to continue developing.

One oocyte survives this process and gets to mature. (Occasionally more than one survives, and that's one of the ways we get twins.) This lucky oocyte continues the process of meiosis. The two chromosome pairs that snuggled together so long are finally separated, and that first meiotic division is finished. But the cell doesn't split into two similarly sized egg cells—why would it? We only need to make one egg cell, and egg cells are *big* (compared to other cells, anyway). The cell divides asymmetrically, with one nice big egg cell and one tiny thing that basically just contains the chromosomes that won't be used. This tiny thing is called a polar body.

Just as in sperm, the cell now contains twenty-three doubled chromosomes: maybe a doubled chromosome 1 from Mom, maybe a doubled chromosome 2 from Dad, and so on. (Whether a given chromosome is Mom's version or Dad's version is the result of chance. It easily could have gone the other way.)

HOW YOU WERE MADE

(What follows is a description of the old-fashioned way. If you were conceived with reproductive technology, such as in vitro fertilization, a few of the supporting details will be different.)

Now the cell, which we can call a secondary oocyte, needs to divide one more time to finish the process. It's still in the ovary, by the way. The second cell division begins, and again it will create one large cell and one throwaway polar body. But then *this* process pauses halfway through, as well. It's frozen, just for a short time until it is fertilized, in a stage called **metaphase II**. The chromosomes are still in their X shapes, but there is only one of each pair. Microtubules from each end of the cell have reached out and grabbed onto the kinetochores at the center of each chromosome. They are ready to pull those X-shaped chromosomes apart—but they have to wait.

While division is paused, it's time for **ovulation**. The oocyte is released from the ovary and travels down the fallopian tube. If no sperm are present, that's the end of the story; the oocyte will continue its journey toward the uterus and will be discarded along with the menstrual lining. Better luck next month.

However, if there *are* sperm cells present—the reproductive tract can store them for a few days to increase the chance that they will be—the oocyte can finally finish its division.

How You Got Forty-Six Chromosomes

This step is a little weird, and you don't need to understand it to understand genetics. But if you're curious, here goes: right after sperm and egg meet, the sperm's DNA (twenty-three chromosomes) and the egg's DNA (twenty-three chromosomes) are each in their own little nucleus. Then each set of DNA duplicates *again*, so they each have two identical copies of each chromosome. These little nuclei don't fuse yet, though! The nuclear membranes fall apart, and the cell divides, pulling forty-six chromosomes to each side. Now we have a two-cell zygote, and the nuclear membranes re-form in each of the two cells.

Finally your cells have the usual number and configuration of chromosomes (forty-six total, twenty-three from each parent), and they can use mitosis to divide over and over and over again to build your entire wonderful self.

When sperm and egg meet, typically in the fallopian tube, the sperm's nucleus enters the oocyte. The oocyte finishes meiosis (producing another throwaway polar body), so that it finally has twenty-three single chromosomes to match the set of twenty-three the sperm carries.

SEX CHROMOSOMES

XX, XY, and Some Other Options

We typically have twenty-two pairs of chromosomes called auto-somes: those are the ones with names like chromosome 1 and chro-mosome 2. But that twenty-third pair is a little different from the rest. They are your sex chromosomes, and in most of us they are either an X and a Y (male) or two Xs (female).

Most of the time, someone with an X and a Y develops male anatomy, including a penis and testes. Most of the time, their gender is masculine—in other words, they identify as a man. Similarly, a person with two X chromosomes will usually have female anatomy (vagina, uterus, breasts) and in most cases will identify as a woman. There are exceptions to these rules, and we'll learn about some of them here.

Your chromosomes and your anatomy are both considered to be part of your **sex**. On the other hand, your **gender** describes who you function as in society and who you feel yourself to be. In this section, we'll just be talking about sex.

By the way, chromosomes don't always perfectly match anatomy. In relatively rare cases, a person can have male chromosomes but female anatomy at birth, or vice versa. A person's anatomy can also be intersex, appearing intermediate between male and female. Chromosomal sex can also occur in configurations other than the typical XX or XY.

THE Y CHROMOSOME'S SPECIAL JOB

The Y chromosome is our very smallest one. It contains fewer than two hundred genes, compared to around a thousand on its partner, the X chromosome. Three hundred million years ago it looked a lot like the X chromosome, but over the years it has shrunk.

A few of its genes are important, and during meiosis the Y chromosome matches up with a small part of the X chromosome that contains similar genes. But the Y chromosome's trademark function is to carry the SRY gene. SRY is named because it is the *sex* determining *region* of the *Y* chromosome. It's like a switch that turns on maleness. If you have an X and a Y chromosome, but your SRY gene is busted, you may not be able to develop male anatomy.

Swyer Syndrome

Someone who has X and Y chromosomes, but a nonfunctional version of the SRY gene, can develop Swyer syndrome. Since they are not able to trigger the production of male anatomy, they end up with a uterus and fallopian tubes, but typically have nonfunctional gonads in place of ovaries. These people are usually assigned female at birth, and grow up to identify as women. Because they do not have functional ovaries, they can't produce egg cells, and need hormone replacement therapy to develop female characteristics at puberty. But because they *do* have a functional uterus, they may be able to become pregnant with a donated embryo.

The SRY gene encodes a transcription factor, a protein that binds to DNA to help turn on another gene. The gene it turns on is called SOX-9, and SOX-9 then turns on genes that are responsible for developing the seminiferous tubules of the testes and the cells inside the testes that become sperm.

Besides SRY and a handful of essential genes that it shares with the X chromosome, the Y chromosome doesn't have much going on.

Scientists believe this is because the SRY gene doesn't match up well with its counterpart on the X chromosome. In creatures with two copies of each chromosome—including humans—we use the second copy as a backup in case of damage. This means if a woman has two X chromosomes, and one of them gets damaged (for example, by a virus or a toxin carried in from cigarette smoke or even from the ultraviolet rays in sunlight), a repair kit of proteins gets to work. They compare the broken part of the DNA to its counterpart on the *other* X chromosome, and are able to fix the damage.

However, if some part of the Y chromosome gets damaged, there is no counterpart on the X chromosome to act as its template for repairs. Consequently, over the years, pieces of the Y chromosome got broken and couldn't be properly repaired. Today, there's very little of the Y chromosome left—but fortunately it has just enough to do its job.

BEYOND XX AND XY

There's another special thing about our sex chromosomes: a person can have more or fewer than the usual number, sometimes without knowing it. Here are some of the possibilities:

- **X:** Turner syndrome. People with just one X chromosome, and no Y, develop as female but cannot go through puberty without hormone treatment. They often have problems with their heart and internal organs.
- **XXY:** Klinefelter syndrome. Because people with Klinefelter syndrome have a Y chromosome, they develop as male. But they don't produce as much testosterone as an XY male. They are typically infertile and may develop breasts. It's also possible to

have more X chromosomes, for example, to be XXXY. The result is similar, but with even less testosterone and a greater risk of learning disabilities and health issues. Only one X chromosome is usually active in each cell; the extra one gets turned off, just like in XX women.

- **XXX:** Triple X syndrome. People with this syndrome develop as female, and simply inactivate two X chromosomes in each of their cells. They may have some of the same learning disabilities and health issues as XXY males, but they tend to have a fully functional female reproductive system.

XX AND XY ONLY DETERMINE SEX IN (MOST) MAMMALS

So far, we've really just been talking about humans. Other mammals, like cats, use X and Y chromosomes to determine sex as well. But when you get into the rest of the world's creatures, sex determination happens in all kinds of different ways.

Birds have a system that is the opposite of ours. Males have two of the same sex chromosomes, while females have one of each. To avoid getting confused, we don't call their chromosomes X and Y; instead, males are ZZ and females are ZW. Some reptiles and some insects also have a ZW system for sex determination.

In another system, used by some insects and a few rodents, females are XX while males just have a single X chromosome; there is no Y.

Bees and ants have an entirely different system: while we humans are all diploid (we have two copies of every gene), bees are only diploid if they are female. A queen can lay eggs that only have one copy of each chromosome, and they develop all by themselves—no fertilization needed—into males. These males have no fathers, and they can't have sons, either. But they can mate with the queen to help her make diploid eggs that will hatch into females.

Some creatures, including some reptiles, don't bother using chromosomes to determine sex at all. In turtles, for example, sex depends on the temperature at which the eggs were incubated. In a pile of sea turtle eggs, buried under the sand, the warmest and the coolest eggs will develop into males; those at an intermediate temperature will become females.

These aren't even all of the sex determination schemes in the animal kingdom. Some animals can change their sex throughout their life. In a population of all-female clownfish, for example, one will change to become a male. In some animals, like the platypus, scientists still aren't sure exactly how sex determination happens.

WHY TWO COPIES?

Diploid and Haploid

Humans have two copies of most of our chromosomes; it's just the way we're built. But it doesn't have to be that way.

We, and other creatures with two copies, are **diploid**. Life-forms that only have one of each chromosome are called **haploid**. For example, moss—which may look like a green fuzzy carpet but is actually made of tiny plants—is haploid during most of its life cycle.

Our two copies give us some advantages. Since we have two copies, we only hand down half to each of our children. Because of recombination and because of the way meiosis distributes chromosomes between the daughter cells, each of our children has a *different* combination of chromosomes from each of us. Perhaps you'll hand your daughter a gene that gives her brown eyes, while you give your son the other copy of that chromosome, and he ends up with blue eyes. Our doubled chromosomes result in more variation among our children.

Plenty of creatures have haploid genomes most of the time, but can also exist in a diploid form that can reproduce sexually (by joining with another individual). Mushrooms are like that; so are their cousins, yeast. We already saw in the "Sex Chromosomes" section that male bees are haploid, but females are diploid.

MULTIPLE COPIES

There are tons of plants with unusual numbers of chromosomes. The number of chromosomes is called **ploidy**. Bananas, for example, are

triploid; they have three of each chromosome, instead of just one or two. Three chromosomes won't split evenly across cells in meiosis, so bananas can't reproduce. You know those brown specks in the middle of your banana slices? Those are undeveloped seeds. The banana's diploid ancestor has giant seeds that make it hard to scoop out and eat the banana flesh, so when this triploid banana happened—probably through some evolutionary accident—people decided to grow more of it. Since they can't grow seeds, banana plants are propagated by removing a shoot or a piece of a stem from the banana plant and planting that on its own.

Ploidy versus Somy

When we talk about a full set of chromosomes being duplicated, we use words ending in *-ploid* or *-ploidy*. For example, our body cells are diploid because they have two copies of every chromosome. On the other hand, if we're talking about the number of copies of one particular chromosome, we use words that end in *-somy*. For example, if someone only has one X chromosome, they have a monosomy, and we can be specific by calling it monosomy X. Or if someone has Down syndrome, caused by three copies of chromosome 21, they have a trisomy called trisomy 21.

Garden strawberries are **octoploid**, meaning they have eight of each chromosome. More DNA means more expressed genes, so this is part of why garden strawberries are bigger than their wild cousins.

X INACTIVATION

Everybody has at least one X chromosome, so it's fine to have important genes on this chromosome; nobody will have to do without. But half of the population only has one X chromosome, while others have two. Therefore this chromosome is in the odd position of needing to carry genes that are fine in either a single or a double dose.

Think for a minute about your autosomes, the chromosomes that just about everybody has two copies of. If it's time to turn on a gene that lives on chromosome 17, both of your chromosome 17s can transcribe that gene. But if something is on the X chromosome, and you're a male with an X and a Y, then you only have one copy to work with.

Our body does have a solution to this problem: just make more of everything on the X. If the genes are transcribed twice as often, you don't actually need two copies.

But what about females and their two X chromosomes? Won't they have double everything they need? Not if they turn off one entire chromosome.

Early in a female's development, something weird happens in each of her cells. Remember how DNA can coil up tightly for storage? One of the X chromosomes in each cell gets packed up, and it won't be used in the cell's day-to-day business from then on.

This happens when an embryo is only about one hundred cells big, and it's a toss-up which of the two X chromosomes gets inactivated in each cell. In cats, there is a gene on the X chromosome that can determine whether the cat produces a dark brown or an orangey color in its fur. Male cats only have one X chromosome, so they can be either orange or black. But if a female cat has the orange and the brown genes on each of her X chromosomes, her coat will be a patchwork of orange and brown according to which X chromosome is still active in each cell.

There are a few genes on the end of the X chromosome that don't get inactivated—those are the ones that are also present on the Y chromosome. So whether you're XX or XY, you still have two active copies of those genes.

HOW WE INHERIT OUR TRAITS

Who Do You Take After?

Be honest: you probably didn't pick up a book about genetics because you wanted to learn about the biochemistry of DNA. (Although if you did, you're my kind of nerd, and we should sit together at parties.) You probably want to know about your own genetics and how you inherited the traits you have.

What's a Trait?

A trait is something you can observe in an organism (such as yourself). Your hair color is a trait. Whether or not you have diabetes is a trait. Traits are often genetic, but sometimes genes are only a small part of what determines if you have that trait or not. In the case of diabetes, your genetics play a role, but so do your diet, your exercise habits, and more.

If you share a trait with one or both of your parents, it seems reasonable to think you got it from them. Biologically, that is—through your DNA.

For a lot of traits, that may be correct. But not all traits come from genes. There are other things that influence our traits. For example, tall parents have tall children, but height comes partly from genetics and partly from factors like how well nourished you were as a child. If you grow up under starvation conditions, you'll be shorter than if you grow up with the exact same genes in a place where you have as much food as you need.

There are also traits like asthma or diabetes where your genes can put you more or less at risk for the trait, but you can also manage your symptoms through diet, exercise, and medication. There's also an element of chance: of a pair of twins with the exact same DNA, one might develop the condition while the other doesn't.

PHENOTYPE AND GENOTYPE

Geneticists have words to describe what you have in your DNA versus what you experience in your body. Your **genotype** describes what versions of a gene you have in your DNA. But the traits that you can see or experience are your **phenotype**.

For example, maybe you have brown eyes. That's your phenotype. However, we can talk about one of the genes involved in eye color and say that you have a certain genotype, with maybe one working and one nonfunctional version of a gene that puts brown pigment into the eyes.

Two people with different genotypes can have the same phenotype. For example, the person in our brown eyes example has two different versions of that gene, which for the moment we'll call the brown and the blue versions. But somebody else who has brown eyes might have *two* working copies of that gene for brown pigment. These two people have the same phenotype, but got it with different genotypes.

On the flip side, the same genotype can lead to different phenotypes. For instance, perhaps there are twins who have the same genetic risk for diabetes, but one develops it and one doesn't. The difference could be diet, nutrition, or chance—or even the effects of *other* genes besides the one we were looking at. Perhaps one twin has a mutation, all his own, that affects a transcription factor for the gene in question.

VERSIONS OF A GENE

We have two of each chromosome (except for the sex chromosomes, which are inherited differently; see the section titled "Sex Chromosomes"). Each pair has a chromosome from each of our two genetic parents. That means that for any gene, you have two versions of it: the one from Mom and the one from Dad.

Mom and Dad

I use the shorthand of a "chromosome from Mom" and a "chromosome from Dad" to help keep track of genes or chromosomes in this book's examples. Many other writers and scientists do the same. But in reality, your parents might not be called "Mom" and "Dad." The people you call your parents might not be the people who contributed the sperm and the egg that became you. (For example, you might be adopted, or perhaps you were conceived with the help of a sperm donor or a surrogate.) It would be more accurate to say you got one chromosome from the person who contributed a sperm to you, whoever exactly that was, and one from the person who contributed an egg.

We have a special name for different versions of a gene: they are different **alleles**. In our eye color example, the person had one allele for a functional brown pigment, and one allele that did not produce the pigment.

Anytime we are talking about a gene that appears on an autosome (one of your numbered chromosomes), we are talking about your two alleles for that gene. Remember, the gene is just the DNA that makes a protein or an RNA. Even though you'll hear people saying you "have a gene" for something, it's not correct. *Everybody* has that gene. However, you can have certain alleles—perhaps you have a mutation in one allele of the BRCA2 gene that affects your risk for breast cancer. That's the correct way to phrase it; you wouldn't say you "have the BRCA2 gene," because everybody, regardless of their risk, has one of those.

HOW GENES BECOME TRAITS

Remember how a gene or "recipe" on DNA gets transcribed into RNA? And from there, the RNA either has a job or gets translated into a protein that has a job? That's what happens with both of your alleles. If one makes a functional protein and the other doesn't, you can still enjoy the effects of the functional protein.

What's a Locus?

Locus (plural: loci) is related to the word location. It means a specific place in a DNA molecule. We often use this word in place of the word gene because it's more precise. Maybe a certain mutation is in a gene's promoter; would you call that part of the gene? (Some geneticists would, and some wouldn't.) The boundaries of a gene are arguable, but a mutation's location is not. It is where it is.

On the other hand, sometimes you'll have two of the same alleles, and then whatever that allele makes will determine your phenotype.

Let's look at blood type as an example. There's a gene called the ABO gene, and at that gene's spot on the chromosome—its **locus**—there are three common alleles.

Each allele of the ABO gene makes a different protein. The A allele creates a protein for the surface of blood cells; let's call it the A protein. The B allele, likewise, creates a B protein. The O allele is the broken one; it does not create any protein.

If you have one B allele and one O allele, your cells will have the B protein. Your genotype is BO, and your phenotype is type B blood. If you have two B alleles, your genotype is BB, but your phenotype is the same as previously mentioned: type B blood. (Likewise, type A blood is a phenotype resulting from AO or AA genotypes.)

If you have both the A and B alleles, you'll have AB as your genotype, and type AB as your phenotype too. If you have two O alleles, you'll have OO as your genotype, and type O blood.

This was a slightly unusual example because there are three possible alleles rather than two. But it's totally normal for a gene to have many different alleles in a population. Any individual can only have two (in diploid species like humans, anyway), but there might be other options that exist somewhere else.

DOMINANT AND RECESSIVE

Finally, We Meet Mendel

When we have two alleles at a certain locus, it can get confusing trying to remember what phenotypes we get from the different ways of combining those two loci. One shorthand is referring to the different alleles as being "dominant" or "recessive." Not all genes can be categorized into two neat alleles, but if they can, it helps us to understand what's going on when a trait seems to skip a generation.

We owe this simplification to a monk named Gregor Mendel, who used pea plants to figure out the difference between genotype and phenotype. He understood—you could say *invented*—the concept of alleles. He did his work in 1866, long before DNA was discovered and its role in heredity was confirmed. It's pretty amazing when you realize he had no idea of DNA, genes, or chromosomes. He just worked backward from the phenotypes he saw and concluded that there must be two of *something* that each pea plant has. These two *somethings* determine what color a pea plant's flowers will be, or whether the plant will be tall or short.

What Is Heredity?

The study of how we inherit traits is called *heredity*. It's related to the word *inherit*: we inherit these traits from our ancestors. If you say a trait "runs in the family," you're talking about its heredity.

Gregor Mendel was a friar, trained as an educator and a scientist, who studied how pea plants inherit their traits. Pea flowers usually pollinate themselves (imagine if you could produce sperm and eggs and make babies all on your own!), so because of the flower's structure Mendel knew pollen wasn't coming in from other sources to mess up his experiments. He also noticed that pea plants come in different varieties, with varying characteristics that he could study.

Why Peas?

Mendel also tried to study genetics by mating mice, but his boss thought it was weird to study animal sex and made Mendel stop. He also studied another type of plant, hawkweed, that didn't give reliable results because it reproduces in a different way than peas. He published the results that worked, and essentially lucked out that pea plants' genetics behave so nicely.

Even though peas usually fertilize themselves, it's possible to open up a flower and carefully transfer pollen (what plants have instead of sperm) onto another flower's female parts to see what happens when *this* pea plant mates with *that* one. He did experiments this way for eight years, carefully observing what offspring resulted from each cross.

HYBRIDS

Mendel studied pairs of traits. In each case, when you take purebred plants with each characteristic and breed them together, you'll find that all of the offspring, or **hybrids**, have just one parent's trait. The traits were:

- White flowers or purple flowers (hybrids have purple flowers)
- Axial or terminal flowers (that is, whether flowers appear on the sides of the plant or only at its tips; hybrids have axial flowers)
- Tall or dwarf overall height (hybrids are tall)
- Round or wrinkled seeds (hybrids have round seeds)
- Green or yellow seeds (hybrid seeds are yellow)
- Constricted (bumpy) or inflated (smooth) seed pods (hybrids have inflated pods)
- Green or yellow pods (hybrids have green pods)

Whichever trait showed up in the hybrids, Mendel called **dominant**. If he crossed white-flowered peas with purple-flowered peas and got a whole field of purple-flowered pea plants, he said that purple flowers were dominant. He didn't know *why* that trait was the one that showed up, but he kept careful notes and was able to tell that one consistently hid the presence of the other. The other trait—in this example, white flowers—he called **recessive**.

Mendel could only study phenotypes; he couldn't sequence the pea plants' DNA. But based on the results from breeding the pea plants, he could tell that there must be two of something that each plant inherits. Luckily, each of the traits he studied was the result of a single locus, and most of the loci were on different chromosomes rather than linked together by being near each other on the same chromosome.

WHY AN ALLELE CAN APPEAR
TO BE "DOMINANT"

We've already seen that genotypes lead to phenotypes because of what the genes involved actually do. For example, if your blood type is A, that's because you have at least one gene that produces the A protein on your red blood cells.

Each of Mendel's traits has a similar story. Purple flowers are "dominant" over white flowers because the purple flowers know how to produce the purple pigment, and the white flowers do not.

Here's an example using flowers:

If the male parent has purple flowers and comes from a purebred line of plants, it will have two alleles that make a purple pigment. Mendel used capital and lowercase letters, so in his tradition the purple-flowered plant would have a genotype of PP.

And let's say the female parent, also a purebred, has white flowers. That parent's nonfunctional pigment gene is designated with a lowercase "p," so its genotype is pp.

When the two parents mate, the male in our example makes haploid pollen that only carry one of the flower color alleles each. Each grain of pollen carries a P. Meanwhile, the female parent can only make ovules (female gametes) with the p alleles that it carries. That means that every offspring from this couple has the genotype Pp.

Since a plant with the Pp genotype can still make the purple pigment, these plants will have purple flowers. We call a genotype with these mismatched alleles **heterozygous**, from the prefix *hetero-*, meaning "different." These heterozygous plants have the same phenotype, but a different genotype, from their PP parent. We can call that parent **homozygous**, meaning it had two of the same allele.

BUT WAIT, THERE'S MORE

Mendel didn't stop there, and neither should we. That homozygous plant, with the Pp genotype, will fertilize itself if we leave it alone. Its offspring will all have a Pp genotype as their male parent (contributing a gamete with either P or p), and they will likewise have a Pp genotype as their female parent. (Remember, this plant is fertilizing itself, so the same plant is contributing the male and the female gametes.)

That means there are three possible combinations for the offspring. They could be Pp like their parent (about 50 percent of them), or they could end up with PP (25 percent) or pp (another 25 percent). And in fact, Mendel could tell from a pea plant's progeny what the parent's genotype was. If the offspring carry a trait in a 3:1 ratio (like our flower example, where 25 percent have white flowers), he knew that the parent was heterozygous for the locus in question.

In genetics classes, students often learn to draw small charts called **Punnett squares** to predict the genotype of offspring from a given pair of pea plants. Punnett squares can also be used to work out crosses for other diploid organisms—like mice, for example.

So even though genes appear to be "dominant" if they show through in the phenotype, there's nothing going on behind the scenes that makes one gene "dominate" another. *Dominant* and *recessive* are just words that describe the pattern of offspring we see.

X-LINKED TRAITS

Why More Men Than Women Are Color Blind

Here's a paradox about the X chromosome: females have more X chromosomes than males, but if a health condition comes from a gene on the X chromosome, males are the ones more likely to show the phenotype.

Don't be confused—it actually makes sense when you remember how the X chromosome is inherited. People who are genetically female have two X chromosomes. Males are XY, so they only have one X.

If you are female and have a recessive allele—say, one that results in color blindness—it will act just like the genes on your other chromosomes. You have two copies, so if one allele makes a functional protein and the other does not, you'll still get the results of having the functional protein.

But if you are male and inherit that recessive allele, you don't have a second X chromosome to rely on. If your one and only X chromosome makes a nonfunctional photoreceptor in your eye, that's it—you're color blind.

WHAT IS COLOR BLINDNESS, ANYWAY?

Many genes affect color vision in humans, and researchers have found at least fifty-six where a mutation can lead to problems perceiving colors. For this example, we're going to look at one of the more common mutations that causes color blindness, or, to use the more accurate term, color vision deficiency. This mutation happens in a gene on the X chromosome.

When light hits our eyes, we only "see" it if it hits a special type of cell at the back of the eye. (That's why we have a pupil and a lens: the lens focuses light through the pupil to project an image on the back of the inside surface of our eyeballs. That back surface is called the retina.)

We have four types of these cells. One type, the rod cell, just registers whether light hits it or not, and zaps a signal down the optic nerve to the brain. The brain can perceive an image based on which rod cells got turned on and which didn't. Think of this as a black-and-white photo: some pixels are on and appear white; others are off and appear dark. We use this type of vision a lot in the dark. Take a look around the room next time you're falling asleep in a dark room. You can't discern any colors, can you? As far as our brain knows, the dark is a black-and-white world.

What Colors Can You See?

Humans have three colors of cones, but other animals don't see the same way we do. Most mammals have just two: one blue and one yellowish. If you've heard that dogs are "color blind," that's true in the same sense that people are often color blind. They will have trouble distinguishing red from green, but they do still see in color. Meanwhile, other creatures have different light receptors than ours. Bees can see ultraviolet light, for example, which is why flowers sometimes have spots and patterns that are invisible to our eyes—but they're obvious to the bees!

But in the daylight, the rod cells all register light, so they're not helpful in putting together a picture. Fortunately, our cone cells are up for the job. We have three types of cone cells. Most of us, anyway.

Some send their signals when they are hit by red light, others when they are hit by blue light, and the third type when they are hit by green light.

Color vision deficiencies occur when one of those three cones doesn't do its job. The two most common types of color blindness, each inherited by 1 percent of male (XY) humans, come from mutations on the X chromosome affecting the instructions for the light-sensitive proteins, or **opsins**, in either the green or red cones.

Let's take the green one, for example. This is the gene that most other mammals, like dogs, do not have. The gene that makes opsin for green cones is known as OPN1MW. Right next to it on the X chromosome is the gene for the red opsins, OPN1LW. We may actually have more than one copy of each of these genes, just lined up next to each other.

When the X chromosomes stick together during meiosis, it's a little too easy for the red and green opsin genes to get mixed up. The recombination process normally just swaps DNA between chromosomes, but if there are near-identical sequences too close to each other, it's possible that they don't match up correctly. It's like when you button your shirt wrong: the buttonhole right below the correct one is still a buttonhole, so you can end up putting all the buttons in the wrong holes before you realize what you've done.

When this happens to chromosomes in recombination, they can end up swapping the wrong amount of genetic material. The result can be that one chromosome is just a little too short or too long. That means the red or the green opsin genes can end up getting deleted in one of the resulting cells. Or if recombination occurs in the middle of a gene, you could end up with a hybrid protein that detects a color that is somewhere in between red and green. (As far as our eyes and brain are concerned, that color would appear as yellow.)

MEN OFTEN GET COLOR VISION DEFICIENCY FROM THEIR MOTHERS

A man with XY chromosomes, who is missing an opsin gene on his X chromosome, will have a color vision deficiency. If his green cone doesn't work, he won't be able to see green. (His red cone will still pick up some of the green light, though, since those colors are actually fairly similar wavelengths of light.)

But since he's an XY male, he gets his X chromosome from the parent that provided the egg cell. Think about it: the only place he could get a Y chromosome is from his father, so the X must come from his mother.

That means his mother had at least one X chromosome with a bad green opsin gene. On the other hand, the odds are, she probably also has a fully functional green opsin gene on her other X chromosome. An XX woman could only be color blind if she inherited *two* X chromosomes with the same nonfunctional gene. That's just less likely to happen.

This pattern holds true for any other X-linked recessive gene, including hemophilia. It doesn't matter if a man's father is color blind, since an XY father can't pass an X chromosome to his son. (If he did, the resulting embryo would be XX and thus female.)

To figure out if a man is likely to be color blind, don't look at his father; look at his mother's other sons (who have a 50/50 chance of getting her affected X chromosome), or his mother's mother's sons. A woman who carries an affected X chromosome will see the trait show up in, on average, half of her sons.

MITOCHONDRIAL DNA

That Other Chromosome

We've spent a lot of time talking about the chromosomes that reside in the nucleus of your cells: in other words, your nuclear DNA. But you actually have another source of DNA inside every cell: mitochondria. If you were a plant, you'd have two: mitochondria and their photosynthesizing counterparts, chloroplasts. We'll focus on the mitochondrial DNA in this section.

Let's zoom out just a little to take a look at our cells. Inside each one—whether it's a bone cell, a muscle cell, a skin cell, or a brain cell, just to name a few—we have a nucleus that contains the DNA we've been talking about. As well, there are other structures in the cell, most of them enclosed in their own membranes, just like the cell itself is held together by a membrane.

The mitochondria (singular: mitochondrion) provide energy to the rest of the cell. Think of them like little generators: you can shovel in fuel, and they give you back energy you can use to power just about anything. When you talk about burning the calories from your food, you're really talking about what these little powerhouses do. They can use carbohydrates, proteins, or fats as fuel, and they supply the rest of the cell with energy-rich molecules called ATP. The ATP, in turn, can give muscles the energy they need to contract, or enzymes the energy they need to build more parts of you.

We only have one nucleus in each of our cells (for the most part), but mitochondria have no such limit. It's not unusual to find cells with hundreds of mitochondria, or even thousands. Sperm pack light for their trip through the female reproductive tract: they only have about one hundred mitochondria. But egg cells are enormous—at about 0.1 millimeters long, they are just barely large enough to be visible to the naked eye. They're so big because they need to contain the raw materials for making an embryo, for its first few days at least. And so the egg contains a whopping 100,000 mitochondria (or more!) alongside its single nucleus.

Burning Calories

Mitochondria convert food to energy with the use of oxygen, so we call their job *aerobic respiration*. If you squat down and pick up something heavy—half of your friend's couch, let's say—your muscles will need a sudden burst of energy. They can actually get that energy *anaerobically*, without the mitochondria's help. But if you go for a run, or even a walk, or even if you just lie in bed and sleep, most of your calories will be burned in the mitochondria, with the help of oxygen. The mitochondria can use almost anything we eat as fuel—sugar, carbs, proteins, fats—as long as other parts of the cell have prepared tiny chunks of those nutrients to feed to the mitochondria.

If you think of nuclear DNA as being truly *you*, then it might help to think of mitochondria as your tiny pets. They have minds of their own—or, to be accurate, DNA of their own. They're part of the family too: mitochondria probably started out as independent creatures, which our cells swallowed and somehow didn't kill. But we've lived with our mitochondria for so long that we've actually swapped some DNA; our nuclear DNA includes genes that mitochondria need to survive and do their job. The mitochondria are here to stay.

THANKS, MOM

Our mitochondrial DNA comes entirely from one parent: the one who contributed the egg cell. The egg attacks and destroys the sperm's mitochondria during fertilization, and so the sperm probably doesn't contribute any mitochondria to the embryo at all. (It's possible that a tiny bit of the sperm's mitochondrial DNA might survive this attack, but if so, it's only a tiny percentage of the mitochondrial DNA in your cells.)

Once that initial fertilized egg begins to divide, those 100,000 mitochondria end up in the resulting cells. Mitochondria reproduce themselves by dividing. If you're thinking of them as pets, imagine that you have 100,000 cats and routinely give half of them away, but the ones that remain are always having kittens, so your house is always full.

Every cell in your body today, if it has mitochondria—and almost all of them do—contains descendants of the mitochondria that your mother gave you. You have the same mitochondrial DNA that she does too.

WHAT'S IN MITOCHONDRIAL DNA?

Mitochondrial DNA is a circular chromosome—yep, a complete loop—about 16,000 base pairs long. It includes thirty-seven different genes, or sets of instructions for RNAs and proteins. Thirteen of these make proteins that are involved in respiration, or calorie burning. The other genes in the mitochondrial chromosome make ribosomal RNAs and transfer RNAs that are needed to make the proteins. (See earlier sections for more on how proteins are made.)

There are no introns in human mitochondrial DNA. There is only one place on each strand where transcription starts, so when the DNA is transcribed, the result is a massive piece of RNA, which is then cut into pieces. The tRNAs are different than the ones in the main part of our cells, so they actually have a different code than ours. For example, if an mRNA made from our nuclear (regular) DNA contains the codon AUA, that matches up with a tRNA that carries the amino acid isoleucine. But the same codon in mitochondria would attract a tRNA that carries methionine, instead. The stop codon UGA is another one that's different; in mitochondria, it doesn't mean "stop." It means "add a tryptophan and keep going."

Mitochondrial tRNAs also allow for more "wobble," which means imperfect matching between a codon and its tRNA. It turns out that most mutations in the third letter of a codon (like the A in UGA) don't really matter. This is a little bit true with human tRNAs, and much more so in the mitochondria.

MITOCHONDRIAL MUTATIONS

If something goes wrong in mitochondrial DNA, a person with that mutation can have a health condition as a result. This would be a genetic disease, but it's not related to any of the genes on our autosomes or sex chromosomes. You would only expect to inherit a condition like this from your mother, since an embryo gets nearly all of its mitochondria from its mother.

Severity of a mitochondrial disease also depends on how many of a person's mitochondria contain the affected DNA. Remember, we have hundreds or thousands of mitochondria per cell. They aren't all identical; they're a bunch of individual organisms that each have babies on their own schedule. They *might* all have identical genomes, but then again, you might have some mitochondria with a certain mutation and some without. If you only have a few affected mitochondria, you might not have symptoms of the disease. If you have many, you may have more severe symptoms.

Cytochrome c oxidase deficiency is one example of a mitochondrial disease. Several mitochondrial genes are needed to build a multi-protein complex called cytochrome c oxidase. This large enzyme performs one of the essential steps in converting food to ATP. Someone with a mutation in one of these genes will not be able to make enough cytochrome c oxidase, and the result can be cell death in cells that require a lot of energy, such as heart and brain cells.

Maternally inherited diabetes and deafness (MDD) is another mitochondrial disease. It's caused by mutations in the MT-TL1, MT-TK, or MT-TE genes, which carry the instructions for certain tRNAs. These mutations slow down protein production. This condition causes diabetes because the cells in the pancreas that release insulin depend on mitochondria to help them figure out when to do it. The mitochondria can't respond quickly because they have trouble producing proteins very well. We don't know yet why this condition also causes deafness.

FAMILY TREES AND AUTOSOMAL INHERITANCE PATTERNS

Mom and Dad and Grandma and Grandpa and...

You know how to draw a family tree, don't you? Start at the bottom of the page and draw a circle if you're female or a square if you're male. (If neither applies, or if you're drawing a pedigree for a long-lost family member whose sex you don't know, draw a diamond.) Draw a line up from that, and then another line, going across like a letter "T," connecting your parents. Horizontal lines indicate partnerships that lead to children; vertical lines represent the relationship of offspring to their parents.

Genetic counselors draw a lot of pedigrees to figure out whether somebody's family history is compatible with certain types of genetic disorders. If you're looking at a particular genetic disorder, shade in the symbols for people who are affected by that disorder.

AUTOSOMAL DOMINANT TRAITS

If an allele is inherited as a dominant trait, like Mendel's purple-flowered peas, everyone with that allele will show the associated phenotype. Plenty of genetic disorders are more complicated than this, perhaps only showing symptoms if a person has the allele *and* certain factors in their life, such as their diet or even bad luck. But there are true dominant traits. Achondroplasia is one.

Achondroplasia is the most common form of dwarfism. It results from changes in the FGFR3 gene. This gene makes a receptor (see the section "What Proteins Do") that is embedded in the outside membrane of the cells that form bones. Normally, this receptor detects when a hormone called Fibroblast Growth Factor is present, and sets off a series of reactions within the cell that result in the cell growing and dividing.

When this receptor isn't working properly, the cells that are supposed to turn into bone don't end up doing their job. Instead of having growth plates at the end of each arm and leg bone, people with achondroplasia barely see their bones grow at all. They end up with arms and legs that are much shorter than usual, so they end up with a very short overall height.

This mutation is a dominant one. People who have just one allele for an imperfect FGFR3 gene have dwarfism. They have some amount of the normal protein but not enough for their arms and legs to grow to full height. People who have two affected alleles can't grow enough bone, though. Besides short limbs, they also have an underdeveloped rib cage and are typically either stillborn or die shortly after birth.

In a pedigree, a person with a dominantly inherited trait like achondroplasia will be able to give this trait to their genetic children. If they have one copy of the affected gene, chances are half of their children will get it. (This is also the case with achondroplasia.) They may also inherit it from one of their parents, although many people with achondroplasia acquired it as a brand-new mutation that their parents did not have.

If many people in the same family have a dominant trait, you'll see the trait in every generation, passed down from parent to child, over and over again.

AUTOSOMAL RECESSIVE TRAITS

Recessive traits are the ones that seem to "skip a generation." Their pattern of inheritance is not quite as predictable as that, but you will notice that these traits will pop up from time to time, occurring more often in a family than you would expect by chance, but not necessarily showing up every generation. It's possible for a trait to stay hidden for many generations, longer than anyone can remember, and then show up again when somebody marries a partner who just so happened to carry the same trait.

Type O blood is an example of a recessive trait. Remember, if you have at least one gene for the A blood protein, you'll have type A blood. If you have at least one gene for the B protein, you'll have type B blood. (If you have both, you guessed it—you're type AB.) It's only if you have *neither* of those proteins that you can be type O. There is also another blood protein, called Rh, that comes in positive and negative. If you have it, you're Rh positive (you might be type A positive, for example) but if you don't have it, your blood type is Rh negative.

If you're type O negative, what would you expect to see in your pedigree? It's possible both of your parents were the same way, but type O negative is pretty rare—just 6 percent of the population. Even if you, personally, are type O negative, it's likely that your partner won't be. When you have children with that person, your children will inherit your partner's phenotype but will secretly carry one of your genes.

To keep it simple, let's imagine you're O negative and you marry someone who is O positive. You have kids together. They all carry your O negative genes, because those are the only kind you have, so they're the only kind you can pass on. But your partner would have also given their Rh positive alleles to your children. If they were homozygous, with two positive alleles, then all of your children will be heterozygous: one positive allele from your partner and one negative allele from you.

On the other hand, if your partner is heterozygous, that means they have one negative allele and one positive allele. Half of the children you have with this person will have a negative allele from you and a negative allele from your partner: congratulations, you've made an O negative baby. But the other half will have your negative allele and your partner's negative allele, so they will be negative.

Recessive traits can hide in the family tree. An Rh negative allele can be passed down from generation to generation without any of the people who carry it ever mating with another partner who carries the allele. But as soon as somebody does, suddenly a child might be born who has the same trait that family members remember: "Your new baby is O negative? You know your great-grandfather was O negative too." This pattern, where the affected trait seems to disappear for a generation or more before popping up again, is what leads people to say that some traits "skip a generation."

SPECIAL INHERITANCE PATTERNS

X, Y, and Mitochondria

Dominant and recessive describe how some genes are inherited, but there are plenty of other special cases. That's not even including the genes whose expression depends on a person's environment and experiences. In this section, we'll look at three inheritance patterns that work differently from the simple patterns described in the previous section.

Y CHROMOSOME TRAITS

The Y chromosome is a simple one to watch: it's passed down from fathers to sons.

There aren't many conditions that come from mutations on the Y chromosome, just because the Y chromosome is so small it doesn't have many genes to begin with. Y chromosome infertility is one, but if somebody has a severe form of the condition, he may not be able to father children at all, so you won't see a very long pedigree.

This condition comes from mutations in a region of the Y chromosome called AZF, for azoospermia factor. The "zoo" part of the name is related to the word *zoo*, as in a park full of animals. It comes from a Greek word meaning "living thing," and it means there are no living things—no sperm—in the semen. The AZF mutations can affect any of several genes that are needed for normal sperm development.

Some people have mild versions of the condition, though, and they may be able to produce enough sperm to have children. In that case, you'll see that a man with azoospermia will have normal

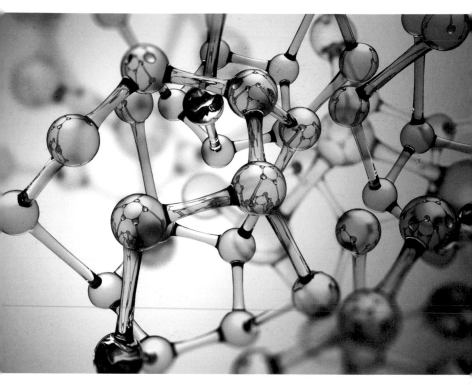

Molecules are collections of atoms bound to one another. For example, when you combine two atoms of hydrogen (symbol: H) with one atom of oxygen (symbol: O), you get a molecule of water: H_2O. Everything in a cell is made of molecules, from the membrane enclosing the cell to the DNA that carries genetic information.

Cell structure

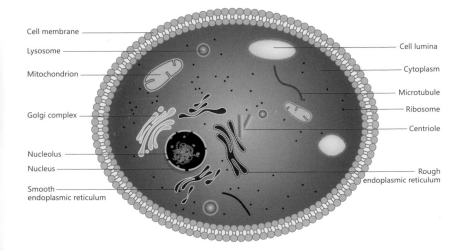

Cell membrane

Lysosome

Mitochondrion

Golgi complex

Nucleolus

Nucleus

Smooth
endoplasmic reticulum

Cell lumina

Cytoplasm

Microtubule

Ribosome

Centriole

Rough
endoplasmic reticulum

This diagram shows the different parts of a cell. One of the most important parts is the nucleus, which contains the coiled strands of DNA that are present in all living things.

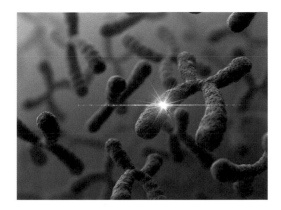

The chromosomes seen here are structures within cells that contain genetic information. Each of these chromosomes has been duplicated into two chromatids, which remain attached at their centromere. When the cell divides, the sister chromatids will be pulled apart, one into each daughter cell.

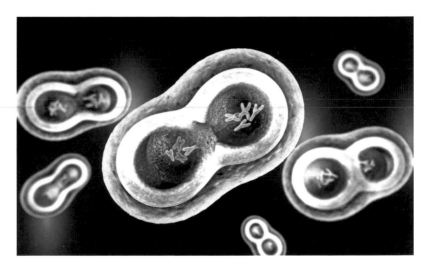

Animal and plant cells divide through a process called mitosis. In this process, the chromosomes in the cell duplicate themselves and move to opposite ends of the cell's nucleus. The cell then separates, creating two cells, each with an identical set of chromosomes. Cell division was first observed in 1835 by Hugo von Mohl (1805–1872), and scientists' understanding of it grew during the nineteenth and twentieth centuries.

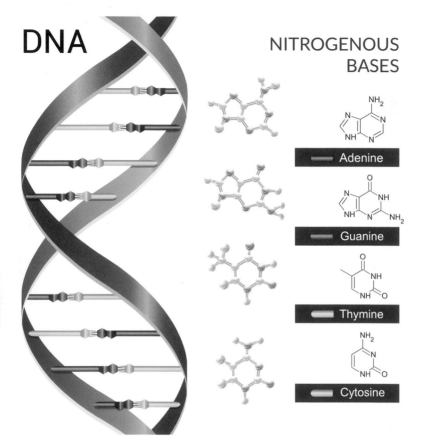

DNA

NITROGENOUS BASES

Adenine

Guanine

Thymine

Cytosine

Deoxyribonucleic acid (DNA) contains the genetic instructions for building a living organism. It consists of two strands, each made up of nucleotides that contain one of four bases (cytosine, guanine, adenine, or thymine) joined to a sugar and a phosphate group. Although the structure seems simple, the order of the bases creates a kind of informational code that contains the genetic instructions for the organism. Bonds between the sugars and phosphates create the backbone of each strand.

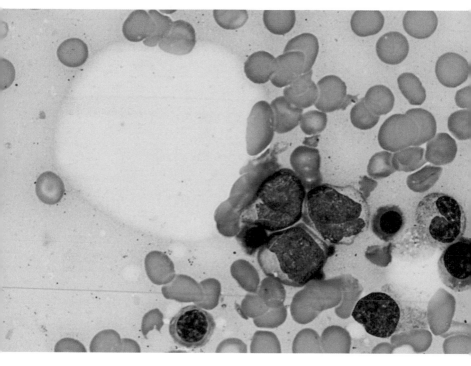

Through the study of DNA and genetics, we've learned a great deal about diagnosing and treating diseases. Scientists study the sequence of nucleotides (the complex organic molecules that make up the strands of DNA) in various organisms. By doing this, they're able to see the nature of harmful mutations, or changes in the molecules. Chief among diseases caused by such mutations is cancer. Here are cancerous cells of a patient with acute myeloid leukemia.

Gregor Mendel (1822–1884), a Moravian friar, founded the science of genetics. By experimenting with peas—crossing short and tall plants, as well as pea flowers of different colors—he discovered that certain characteristics can be passed predictably from one generation to another.

The Empress Alexandra of Russia (1872–1918) carried the gene for hemophilia, a disease that prevents blood clotting. Even though she wasn't a hemophiliac, she passed this disease to her son, Alexei. In an effort to treat the disease, she and her husband, Tsar Nicholas II, turned to a "holy man," Rasputin, who became increasingly unpopular because of the power he wielded at court. He was assassinated in 1916, but his influence over the tsar and tsarina probably contributed to the overthrow of the monarchy the following year.

Francis Crick (1916–2004) and James Watson (b. 1928) discovered that DNA, the most important molecule in all living things, consists of two molecular strands wound about each other in a double helix. Every cell in the human body contains instructions, coded in DNA, on how to build a human being. Watson and Crick's discovery was a tremendous boost for genetics, since it showed how genetic information reproduces itself and where mutations can occur in the process.

James D. Watson.

Rosalind Franklin (1920–1958) was a researcher at King's College London in the 1950s. A highly skilled scientist, she created the clearest set of images of the x-ray diffraction patterns of DNA molecules that had been produced up to that point. Without her knowledge or permission, her supervisor (with whom she was feuding) later showed these images to Watson and Crick, who looked at the images and realized they supported the idea that DNA was shaped like a double helix.

daughters, but his sons will all inherit his Y chromosome and thus should all have azoospermia too. If the man inherited his mutation—rather than being the lucky recipient of a brand-new one all his own—he must have gotten it from his father (or whoever provided the sperm, because this is about inheritance rather than family relations). After all, there is no way his mother could have given him any Y chromosome, since she provided the X in his XY.

X CHROMOSOME TRAITS

X chromosome traits, when they behave like recessive genes, can be very confusing ones to try to track in a pedigree. The affected people will probably all be males, but the inheritance pattern is not father-to-son like with Y chromosome traits.

How Does Blood Clot?

Blood clotting may seem simple, but it's actually a complex process. When we have an injury that makes us bleed, tiny cells in the blood called platelets spring into action, sticking together. They also trigger proteins called fibrinogens to convert into fibrins, which are (you guessed it) fiber-like strings of proteins that can pack together to form a clot to plug the wound and stop any more blood from leaking out.

Remember that females have two X chromosomes, so if they have a recessive trait on one X chromosome, the other is likely to have a healthy copy of the gene, and this person will not show the associated phenotype. We learned about color vision deficiency in the "X-Linked Traits" section; recall that a small percentage of men are color blind, but the condition is extremely rare in women.

Another well-known trait that's carried on the X chromosome is hemophilia. This is a condition in which blood cannot clot, or clots very slowly. Before modern treatments were available, it was often fatal, since boys with the condition could die in infancy or childhood from what would otherwise be a minor injury.

The clotting process requires many different proteins. These proteins can be activated, which often just means that they are able to activate the next protein in the process. Several steps of the process happen all at once, and there's even a feedback loop, where activated proteins go back to the original proteins and activate them *even more*.

One cascade of proteins ends up activating clotting factor VIII, while another cascade activates clotting factor IX. These two proteins then work together to activate the next protein in the series. If you can't properly make factor VIII, you have a condition called hemophilia A. If you can't make factor IX, you have hemophilia B.

Britain's Queen Victoria is probably the most famous carrier of the disease. It may have been in her body that a mutation occurred in the gene for clotting factor IX. (None of her ancestors are known to have had hemophilia.) Queen Victoria herself was fine, but her son Prince Leopold had hemophilia. He died at age thirty of blood loss from a minor fall.

Two of Queen Victoria's daughters, Princess Alice and Princess Beatrice, went on to have sons who also had hemophilia. These two princesses also had daughters, and those daughters also gave birth to sons who had hemophilia. Victoria's granddaughter, Alix, was also a carrier, with the result that her son by Tsar Nicholas II of Russia was a hemophiliac. The royal couple sought the services of a "holy man," Rasputin; thus hemophilia indirectly contributed to the onset of the Russian Revolution.

Here are the features of a family tree with an X-linked, recessive trait:

- Males are usually the only ones affected.
- Affected males have a mother who carries the condition.
- Female carriers of the condition will give it to half of their sons. Half of their daughters will be carriers.
- Males with the condition will give an affected allele to all of their daughters (who will then be carriers).
- Males with the condition cannot give it to their sons, because they give their sons a Y chromosome, not an X.

MITOCHONDRIAL TRAITS

A person with a mitochondrial disease, if they inherited it, got it from their mother. Remember, mitochondrial DNA lives inside tiny organelles in our cells called mitochondria. An egg cell contains over 100,000 mitochondria. These are the mitochondria that the zygote (and thus the embryo, fetus, and baby that it becomes) inherits. Sperm contain far fewer mitochondria, which tend to be destroyed at fertilization.

This means that a female with a mitochondrial disease can pass it down to her children, but a male with a mitochondrial disease cannot.

That said, a female with a mitochondrial disease isn't *guaranteed* to pass it down to her children. The presence and severity of the disease depends on how many mitochondria in each cell contain the affected DNA. If a woman has, let's say, 50 percent of her mito-chondria affected with a certain mutation, it's possible that as her ovaries were forming, some of the cells divided in such a way that the affected 50 percent ended up in one cell and the non-affected 50 percent ended up in another. More likely, each egg cell gets a differ-ent half of the woman's original collection of mitochondria, so that if you looked at a series of her egg cells over time, you would see that some have nearly 100 percent of their mitochondria affected, while others have almost none.

On average, you would guess that you'd see a lot of cells with 50 percent of their mitochondria affected. But for any individual egg cell, and thus any individual child, all bets are off. In a pedigree, you will see that anybody who is affected has a direct maternal line to an XX female who was also affected—likely their mother.

NATURE VERSUS NURTURE

Who Shapes You?

This is where we take a step back and ask, *Why are we looking at our DNA at all?*

Your DNA carries instructions for how to make you, and these sets of instructions, or genes, are sometimes good, sometimes faulty, and sometimes somewhere in between. We've been talking as if this is the whole story, that if there's a problem with your DNA then obviously there will be a problem with you as a human being. But that's not true, and it's time to disentangle what DNA can do from what it can't.

WHY THE UNCERTAINTY?

First off, scientists don't understand 100 percent of how each gene works, especially when you start asking questions like: How much does it matter? When does it get turned on? What happens if you have a gene that increases your risk for something, but then you also engage in behaviors that decrease your risk?

Secondly, genes aren't usually as simple as functional versus broken, or turned off versus turned on.

For example, recall that for a gene to be transcribed into RNA, a group of proteins called transcription factors have to grab on to specific parts of the DNA—maybe right in front of the gene, or in some cases they might need to bind to other areas of the DNA.

If a mutation just affects one single base pair in the DNA sequence that a transcription factor protein needs to bind to, that doesn't mean that gene will never ever be transcribed. However, it might mean that the DNA doesn't match up as strongly to the transcription factor. Perhaps it will bind, but weaker. That means that some percentage of the time, the transcription factor might detach from the DNA before transcription starts. Not always! Just sometimes. The gene will still get transcribed, and it will still be made into a protein (or whatever gene product), but just not quite as often.

Gene expression also tends to be a multifaceted process. Maybe there are several different chains of events that can happen to turn on a gene. If one of those chains of events doesn't happen reliably, but the others do, the gene might not be turned on as often. But that really depends on which pathways are being used, and that might depend on other factors such as what the person is eating, whether they smoke, and more.

Penetrance

If people have the right genes to make a certain trait show up, what percentage of the time do they actually show the trait? That number is the allele's penetrance.

Many genes related to cancer or chronic diseases are not fully penetrant. For example, if you have a mutation in the BRCA1 or BRCA2 genes, you will have a higher risk of developing breast cancer than the rest of the population. But there is no BRCA1 or BRCA2 genotype that guarantees breast cancer. Plenty of people will develop breast cancer who don't have any unusual mutations in those genes.

ENVIRONMENTAL FACTORS

Aspects of your life that *don't* include genetics are often called **environmental factors**. This doesn't necessarily have anything to do with "the environment" in the sense of the natural world, forests, oceans, and so on. Environmental means *your* personal environment. It can include pretty much anything that's not part of your DNA.

For example, if you smoke, that's an environmental factor. If you eat a high-fat diet or a low-fat vegetarian diet, those are environmental factors. If you exercise all the time, or if you get injured and become paralyzed, those are environmental factors too. If you were malnourished as a child—you guessed it.

Environmental factors play a *huge* role in our health. Take obesity, for example. It's absolutely heritable: you can end up with gene variations that make your body more likely to store fat, or to have a difficult time burning it off. There are genes that affect your metabolism and genes that affect how good you are at different types of exercise. Genes in the brain can affect how you feel about your weight and whether you end up developing an eating disorder or not.

But there are also "environmental" factors that affect your weight. How much food you eat is a major one. If you live in a place where high-calorie food is easily available, you'll be more likely to gain weight than if you live in a place where food is scarce. Stress can affect your metabolism and make your body less likely to give up its fat stores. If you work a job that keeps you up at night, that also seems to screw with your metabolism. Whether you develop an eating disorder, and what kind, might depend on the experiences you have during your childhood and your life. If you have access to therapy to help you deal with mental health problems, you might be more likely to avoid an eating disorder. If you have access to a nutritionist and other support when you need to lose weight, those factors can also end up influencing the number on the scale.

SIMPLE EFFECTS

The interaction between nature and nurture—or genetics and environment—can also be a simple one. Siamese cats are usually a pale color, with dark paws, ears, and face. (Rabbits and mice can also have this same color pattern.) But nothing in the cat's genetics dictates that it should have black pigment in one body part versus another. The explanation for this cat's pattern is simple biochemistry.

Black pigment, or melanin, is made with the help of an enzyme called tyrosinase. If a cat has two fully functional tyrosinase genes, it will have colored pigment all over. The tyrosinase gene can also be nonfunctional, which can lead to albinism. Albinism is a recessive condition that results in a cat with no pigment in its skin or fur, or even its eyes. (Albino animals usually have red eyes because they are missing the pigment that would otherwise make the eyes a color such as brown.)

But Siamese cats don't have a fully functional *or* a broken version of the tyrosinase gene. Instead, they have a version that is a little bit temperamental and only does its job at cool temperatures. A cat's paws, ears, and face are not as warm as the rest of the body. (Think about how your fingers get cold in chilly weather and how you can warm them up by holding your hands against your torso.) All of the cat's skin cells make the protein, but the protein is only able to do its job in the right conditions.

You can prove this by keeping animals with this allele, like Himalayan rabbits, in a warm environment. Himalayan rabbits are normally white with black paws, tail, ears, and nose. But if you keep them at temperatures above 86°F, the bunnies grow up without any pigment, and they look the same as albino rabbits.

EPIGENETICS

The Environmental Mark

By now we're used to thinking of DNA as a way of storing information. But the information that's coded in the sequence of nucleotides (A, T, G, C) isn't all. Your DNA also carries **epigenetic** information, which can change while the sequence of bases stays the same.

Epigenetics literally means "on the genome." The cell can make changes by adding methyl groups (a carbon and three hydrogens) to certain parts of the DNA, or it can add other groups of atoms to specific places on the histone proteins that are snuggled up next to specific parts of the DNA.

These changes affect which parts of the DNA get transcribed into RNA, and thus they dictate what genes are expressed. Epigenetic modifications to the DNA are changes in gene expression that persist even when the DNA is replicated. When a cell divides, both copies end up with not just their parent cell's DNA sequence, but also their parent cell's epigenetic modifications: those methyl groups and histone tags that can continue to direct gene expression even after the cell divides.

These changes persist when a cell divides, but that doesn't mean you inherited them from your parents. Before an egg and sperm meet, some of their epigenetic modifications are reset, so you get the chance to come up with your own. But we do know that *some* epigenetic changes seem to be heritable from parents to children. Just be aware that when we talk about epigenetics, that's only one small piece of how they work.

DNA METHYLATION

Specialized enzymes add methyl groups to cytosines in places where the cytosine comes right before a guanine in the DNA sequence. (These places are called CpG sites, because they have a nucleotide with a cytosine, then after the next phosphate on the DNA's backbone they have a guanine. In other words, we're not talking about a cytosine/guanine base pair, but rather a cytosine and guanine next to each other on the same side of the DNA.)

Experiments in mice show that if they don't have the enzymes that add these methyl groups, they can't develop normally in utero. In mammals (including humans and mice), 70 to 80 percent of our CpG sites are methylated. The places where they *aren't* methylated are usually gene promoters. Promoters are necessary to start the transcription of a gene, as we learned in the "Turning Genes On and Off" section. That means that methylating a promoter probably turns it off and is something cells do when they and their offspring aren't likely to need that promoter again. Other places in DNA can be methylated as well, and we don't understand all of the reasons for methylation.

Shortly after egg and sperm meet, the resulting cell takes the methyl groups off most of the methylated areas of DNA, as if hitting a "reset" button. As the embryo develops, genes are methylated again, little by little. We think this process is part of how stem cells, which are able to develop into any type of cell, settle down and specialize into skin cells and other specific types of cells that we need.

HISTONE MODIFICATION

You remember histones, right? They are the proteins that DNA is always wrapped around. Think of them as millions of tiny spools, each carrying one or two loops of the thread we call DNA.

Histones aren't shaped exactly like spools. Each one is made of eight separate proteins: two each of H2A, H2B, H3, and H4. And each of these proteins has a "tail" that sticks out.

There are several places on each of those tails where different chemical groups can be added. For example, an enzyme could come by and add a methyl group to a certain spot. That's the same methyl group we just talked about on DNA ($-CH_3$), but this time it's on the histone protein instead of being on the DNA. Or, instead of a methyl group, an enzyme might add an acetyl group ($-CH_3CO$) or a phosphate ($-PO_4$). Certain histone modifications recruit enzymes that methylate DNA, and vice versa, so these two methods of silencing genes often end up reinforcing each other.

Chromatin Structure in Epigenetics

Histone modifications can also determine how tightly DNA is packed. When the DNA is packed up so tightly it can't be transcribed, that's heterochromatin. The opposite, euchromatin, is DNA that's loosely packed and that the cell is able to use.

Noncoding RNAs also help regulate chromatin structure. These RNAs are transcribed from DNA, but instead of going on to be translated into protein or act as an enzyme, they bind to DNA and cause the cell's machinery to modify histones, methylate DNA, and change the structure of chromatin.

These modifications tell the cell whether the DNA wrapped around that histone is active or not. For example, if there is an acetyl group at the lysine residues of H3 and H4 histones, that chromatin is active. But if there is a methyl group at the ninth lysine residue of H3, and methyl groups on the cytosines in the DNA, that chromatin is inactive. There are more possible configurations, and these are called the histone code. However, different organisms use this code differently, and scientists are still trying to figure out exactly how to read it.

PARENTAL IMPRINTING

Even though we have two copies of all our genes—one from each parent—sometimes we only use one or the other. These genes are **imprinted** by the sperm or the egg cell, so that our cell sees the two copies as different. (This only happens for a small percentage of our genes, perhaps 1 percent, although that number may turn out to be larger as scientists learn more about epigenetics.)

The gene for insulin-like growth factor 2, IGF2, is only expressed in humans from the paternally imprinted allele (in other words, the one you got from your father or whoever provided the sperm). Some genes are expressed differently depending on which parent contributed them. One mutation, in a part of chromosome 15 that encodes the UBE3A gene, causes a disorder called Angelman syndrome if a child inherits it from their mother. But the same variation of the same gene, if inherited from the father, produces an entirely different disorder, with different symptoms, called Prader-Willi.

WHY EPIGENETICS MATTERS

It's pretty cool to know that gene expression is regulated this way. But what's even more intriguing is that we can acquire epigenetic modifications that get passed down to our offspring. This has been studied for babies in utero, for example, in a famous study of Dutch women who were pregnant during the winter of 1944 and 1945. There was a famine that year. Thousands of people died; those who survived had to make it through the winter on four hundred to eight hundred calories per day. Those babies were shorter at birth and developed health problems later in life.

When *those* children had babies of their own, those children—grandchildren of the women who had starved—had some of the same health problems as their parents. Their DNA sequence hadn't changed, but they seemed to carry epigenetic changes that resulted from that famine. Epigenetic changes aren't inherited as reliably as DNA sequence, and scientists are still trying to figure out how they are transmitted from parent to child.

Environmental toxins and chemicals also seem to be able to affect DNA methylation. The same is true of tobacco and alcohol. These chemicals' effects on epigenetics might explain part of why they can increase our risk of cancer. On the flip side, a healthy diet and lack of prenatal stress may be able to contribute to healthier epigenetic patterns.

TRAITS CAUSED BY MANY GENES

Why Stop at Just One?

It's time to talk about the truth behind another of our simple ideas: the idea that a gene makes a protein, and that one protein is responsible for your phenotype, or appearance.

We already learned how our environment and experiences can affect whether a gene is expressed and used, or not, but there's another important thing to know: complex traits are rarely decided by just one gene.

GENE INTERACTION

Gene products don't act alone; they interact with other proteins, RNAs, and other factors. Sometimes we're talking about a literal interaction; imagine a receptor that's designed to match up with a certain other protein. Their ability to bind together depends on *both* of the proteins being the right shape.

Other times, genes "interact" simply by acting in the same or related pathways. One way this can happen is called **epistasis**, where one gene masks the effect of another totally different gene that lives in another place on the genome.

Here's how it works. Think of Labrador retrievers, those hungry, friendly, short-coated dogs that come in black, chocolate, and yellow. There isn't just one gene that determines coat color; there are actually two major ones.

One is a gene called TYRP1, for tyrosinase related protein 1, that makes a black pigment called eumelanin. There are two common alleles of this gene: one where the pigment comes out as black, and another that appears brown. Dogs that have two of the brown version could end up looking brown all over, and in that case they would be called "chocolate" labs. Dogs with at least one black version might be black labs.

Why are we saying "could" and "might"? Because their final color actually depends on *another* gene. This one, melanocortin 1, controls how the brown or black pigment gets into the dog's hair. If this gene isn't fully functional, it doesn't matter what color pigment the dog makes; none of it, or almost none, will end up in the dog's fur. In that case, the dog will be a pale color, and we'll call it a yellow lab.

QUANTITATIVE TRAITS

Most of our traits are complex ones, involving far more than the two genes that contribute to Labrador retrievers' fur color. (Actually, even labs have more genes involved in fur color than just the ones we mentioned previously. For example, there is a third gene that determines whether a yellow lab is a pale or a darker yellow.)

After all, there aren't just two heights of human beings, as you would expect if there was a single gene for height. There *is* a single gene that controls whether a person has achondroplasia (one type of dwarfism), but it acts in addition to other genes that tweak your exact measurement.

How many genes affect height? A lot, it turns out. Scientists have combed through the genomes of people of various heights and found more than seven hundred different variations that seem to be linked to height in some way. For example, if people who have one allele are consistently a smidge taller than people with a different allele at that locus, it goes on scientists' "maybe" list.

Some of the seven hundred variations only have the power to add or subtract a little bit to our height—maybe a millimeter or so. But some can affect our height by a centimeter or more. That doesn't sound like much, since there are 2.54 centimeters in an inch. But in total, all of those genes contribute to the final number that shows up when you stand next to the big ruler at the doctor's office.

And yet, genetics can't explain everything about our height. Environmental factors play a role too. People who were malnourished in childhood grow up to be smaller than they would be if they grew up with plenty of food on hand. People whose immune systems had to deal with frequent infection might be shorter at adulthood too. There are also gene interactions at play: women tend to be shorter than men, likely because of something about their genetics or biology. And since the signals to stop growing are linked with puberty, variations (genetic or otherwise) that affect the timing of puberty can also affect a person's final height.

But even with all of our knowledge about genetics and environmental factors on height, we can't totally predict how tall a baby will grow up to be. The best method is still the one based on phenotype: average the mother's and father's height, and then add 6.5 centimeters if the person is male, or subtract 6.5 centimeters if she is female.

HOW YOU KNOW WHEN YOU'RE DEALING WITH A QUANTITATIVE TRAIT

The simplest way to know a trait is quantitative is simply that you can't easily categorize or explain it genetically. Height obviously runs in the family—tall parents have tall children, for example—but you can't easily divide people into tall and short categories. You also can't draw a pedigree that explains who inherited a tall versus a short gene. (Unless you're looking at a trait like dwarfism or acromegaly, where there are a few single-gene variations that can drastically change a person's height. Those are rare, though.)

For example, height in animals is also a quantitative trait, and so is their weight. If you're looking to breed a bigger chicken or a fatter hog, you can't expect one gene to totally determine how large the animal is going to be. The same goes for a lot of other traits, like how much fatty marbling is in a cow's muscle tissue.

Farmers also look for ways to increase their yield of crops, whether those are tomatoes or grains of corn or anything else. Crop yield is another complex trait, depending on the environment but also on a complex constellation of genes deep inside the plant's DNA.

Skin color in humans is also the result of many different genes that affect how much melanin we make in our skin and how it gets distributed. If somebody with very light skin and somebody with very dark skin have a baby together, their child is likely (but not guaranteed) to have a skin color that is in between the parents', because they inherited some dark gene variants and some light ones. The flip side of that is that when two people with the same skin color have a baby together, the baby might end up with a collection of genes that included their parents' lighter variants, or, on the other hand, the darker ones. It's possible for a family to have children with a variety of different shades of skin tones.

SIMPLE AND NOT-SO-SIMPLE TRAITS

You May Know These

While plenty of traits are inherited in a complicated way—as we learned in the last few sections—there still are plenty of human traits that we usually owe to a single gene. Here are a few you may have heard of or may be able to observe yourself.

LACTASE PERSISTENCE

As babies, nearly everyone is able to drink and digest milk. About 40 percent of the calories in human milk come from a sugar called lactose, so babies are equipped with an enzyme called lactase. (Note the spelling: words ending in -*ose* are sugars; those ending in -*ase* are enzymes.) But as we get older, the lactase gene gets turned off. Nothing wrong with that—after all, it's not like twenty-year-olds need to breastfeed. If you *don't* have lactase, and you drink a lot of milk, you'll feel gassy and bloated, and maybe have diarrhea. It isn't a life-threatening condition, but it's not pleasant.

But somewhere along the line, humans started keeping animals for their meat and milk. If you ferment milk, for example, turning it into yogurt, bacteria grow in the yogurt, making chemical changes to the milk. As a result, the milk becomes more acidic, which prevents other bacteria from growing in it. That means it's a form of food preservation. The bacteria also have their own lactase enzyme, so in a sense they are digesting the lactose for us.

Fermented milk products are popular across the globe. Besides yogurt, there's also kefir and thousands of different types of cheeses. But one day—we think about 7,500 years ago—somebody in Europe was born with an accidental tweak to their genome: their lactase gene never turned off.

That person and their descendants were able to drink milk straight from the cow without worrying about diarrhea. That was a useful enough mutation that people who had it were more likely to survive, or at least had more children, than those who didn't. They could get the extra nutrients from milk without having to take the time to make it into yogurt or cheese. Today, around 90 percent of people from northern Europe can digest lactose as adults, compared to fewer than 5 percent of people from Asia.

Most commonly, lactase persistence is a trait from a single gene, and it is dominant. People with two recessive alleles are sometimes called "lactose intolerant." Even though there is a gene clearly linked to this trait, environment does play a role. The bacteria in your large intestine are what create the symptoms of lactose intolerance as they digest the lactose that your body didn't. The exact balance of bacteria in your body can affect how strongly you feel the symptoms. So, of course, do your eating habits: you can't react badly to milk if it's something you never drink.

EARWAX AND SWEAT ODOR

Is your earwax gooey, sticky, and golden brown? If so, you have "wet" earwax. Or perhaps your ears contain something more crumbly or flaky that has a grayish color. That's "dry" earwax.

The glands that make earwax are a type of apocrine gland, similar to our sweat glands. A gene called ABCC11 makes a protein that gets excreted by this type of gland. If you have gooey earwax, you also have a bonus trait: stinky sweat. (Bacteria eat the secreted ABCC11 protein in our sweat, and it's the bacteria that cause the odor.) If you have dry earwax, your armpits probably don't stink. Lucky you. Dry earwax and non-stinky sweat are most common in east and northeast Asia.

TRAITS THAT AREN'T SO SIMPLE

Amazingly, we don't know the full story of the genetics behind common, obvious traits that we have. We already saw how height is determined by hundreds of genes, for example. Here are a few more human traits whose genetics are surprisingly hard to figure out:

Eye Color

There are a few genes that commonly affect eye color, but they don't tell the whole story. A gene called OCA2 makes a protein that we use to manufacture melanosomes, little nuggets of pigment that can color our eyes. If you have brown eyes, this gene is probably fully functional, and it puts pigment-containing melanosomes into the iris of your eye.

If you don't have brown eyes, that might be because your version of this gene makes a protein that doesn't do the job. Geneticists once thought this was the whole story: either you have brown eyes thanks to one pigment gene, or else you're missing it and you have blue eyes.

There are other genes that affect eye color. For example, you may have a variation in an intron of a nearby gene, HERC2, which regulates whether the OCA2 gene gets expressed. There are at least a dozen other genes that influence eye color. (There are also more eye colors than just blue and brown!)

Hair Color

Quick: how many hair colors are there? If you answered with a small number, you're probably wrong. Sure, we can group hair colors into categories like red, blond, brown, and black. But if you analyze enough samples of hair, you'll find enough variation to conclude that there must be more than a few genes involved.

Black or brown hair contains a pigment called eumelanin. Just as in Labrador retrievers (see "Traits Caused by Many Genes" section), it may be brownish or it may be jet black. Another pigment is pheomelanin, which appears red. These pigments occur in other parts of your body, not just your hair. You can thank eumelanin for brown freckles, and pheomelanin for pink lips. Red hair is full of pheomelanin, with little to no eumelanin. And black or brown hair is packed with eumelanin, although it may also have pheomelanin as well. If you bleach your brown hair, and it turns orange, you may be looking at your natural, previously hidden pheomelanin.

Both pigments are made from an amino acid called tyrosine. Different enzymes—whose instructions are encoded in our genome—convert tyrosine into a slightly different chemical, and then that chemical into yet another chemical. Eumelanin and pheomelanin are two different end products of this pathway. To make pigmented hair, we also have to get that pigment into the hair in the first place, so there are enzymes that do that job too. Variations in any of these enzymes can affect how much of each pigment makes it into our hair, and thus what color our hair appears.

Hair color has other factors too. Adults with brown hair may have had blond hair as children, for example. And many of us will turn gray as we age, because the cells that produce the pigment, called melanocytes, are genetically instructed to die off. So we can't describe hair color in simple terms. Somebody who "has brown hair" may have been born blond, then deposited both brown and red pigments in their hair for an auburn color, and then finally gone gray or white in their old age.

PERSONAL GENOMICS

What Happens When You Spit in a Tube?

Since your DNA holds all kinds of information about you, wouldn't you like to read what it has to say? That's the idea behind personal genomics services. From a sample of saliva, they can tell you about some of the information in your DNA. Here's how that happens.

FIRST, YOU SPIT

A typical DNA testing kit is just a weird-looking tube. Depending on what company you ordered it from, it will probably come in a snazzy-looking box with a set of instructions and a prepaid mailer to return it to the company for analysis. There is also usually a code to enter in a website or app. With this code, the company can connect your name and account with the sample that shows up in their mailbox. It usually takes a while for the DNA to be processed and for you to see your results: four to eight weeks in most cases, but again that depends on the company and on how busy they are at the time of year you send it in.

The tube has a funnel on top, and you spit into this funnel to fill the tube up to a certain line. Usually you need to provide two milliliters of saliva to be sure there's enough DNA to work with, so you'll spend a few minutes spitting into the tube before you can be sure you have enough.

Why saliva? Let's take a quick look at what saliva actually is. Glands in our mouth squirt out the liquid component of saliva, which includes water, electrolytes, and mucus. Its main jobs include moistening the food we eat and keeping our mouth healthy in between meals. Saliva contains enzymes that kill bacteria, for example. This liquid itself doesn't contain DNA. But once this liquid is in the mouth, cells can end up floating around in it. Cells flake or rub off from the inside surface of the cheek. These are called buccal epithelial cells. The mouth also contains white blood cells, best known for their role in our immune system. These cells float in our saliva as well, and make up 0.5 percent of what's in that tube.

When you spit into a tube, you're filling that tube mostly with liquid saliva, but also with those cheek cells and white blood cells that are mixed in. These cells contain what we really want: full sets of our DNA.

The next step, closing the cap, is also an important part of the process. The cap of the tube contains another liquid, called a buffer, that helps to preserve the DNA while it sits at room temperature in your house and in the mail trucks that take it to the lab. (Otherwise, DNA samples are usually kept in a freezer.) The buffer is held in the lid by plastic film, and when you close the lid, the plastic is punctured so the buffer can flow into the tube and mix with your saliva. After that, you'll probably be instructed to shake the tube and replace the funnel top with a more secure cap. Then you drop it in the mail and wait.

EXTRACTING DNA FROM THE SPIT SAMPLE

First, the lab extracts the DNA. Cells contain a lot of stuff, but we only care about the DNA, so we have to break open the cells and get rid of all that other stuff. (We don't care about the enzymes and liquid in the saliva, either.) Soap-like chemicals are able to dissolve the cell membranes, for starters. The next step is usually spinning a tube of the resulting soup in a centrifuge.

A centrifuge looks like a tiny washing machine with slots around the edge of the spinning part. Capped tubes containing the sample machine fit into these slots, and when the machine turns on, the force of the spinning makes the densest parts of the sample fly to the outermost part of the circle, which is the bottom of the tube.

When the centrifuge is done spinning, all the heavy stuff will have stuck to the bottom of the tube, and in fact it ends up as a solid little pellet. The liquid that's swimming on top, called the supernatant (yes, that word literally means "swimming on top"), contains the DNA.

AMPLIFYING DNA

When we amplify sound, we make it louder, so it's easier to hear. When we amplify DNA, we make more of it. That makes it easier to detect the DNA among all the other chemicals in the same test tube. Decades ago, before we had technology to easily amplify DNA, genetic testing required a lot of material. But thanks to amplification, a personal genomics lab can tell you all about your DNA based on the tiny amount—perhaps one hundred micrograms—that's found in the cells in your spit sample. Amplifying is done with the polymerase chain reaction, or PCR, discussed later.

HYBRIDIZING (FOR GENOTYPING SERVICES)

If your personal genomics system is one like 23andMe that works by **genotyping**, the next step is hybridizing your DNA to a glass chip, so called because it looks something like a computer chip. 23andMe's chips are about the size of a stick of gum, and they're covered with microscopic glass beads. Each tiny bead has a little bit of DNA attached, with either a red or green fluorescent molecule. If your DNA sticks to the one with the red label, the computer that scans the chip will see a tiny red dot in that area. If you have a different sequence that matches the DNA with the green one, the computer will see a green dot. If you have both, the two colors will appear together, to the computer, as a yellow dot.

The same situation occurs for millions of beads across the chip. Each dot tells the computer whether your DNA matches the specific sequence of each of the little samples of DNA on the chip. The result is not a sequence that you can read like a book, but rather a collection of yes/no answers for about a million different points across the genome. Each of these points is a place that may have a **single nucleotide variant**. Your results from a genotyping service can tell you whether you have *this* versus *that* version of a certain common mutation. However, since this technique only looks at about a million specific letters out of your entire three billion base pairs of DNA, it can't tell you everything about yourself.

SEQUENCING (FOR EXOME SEQUENCING SERVICES)

Other companies, including Helix, take a different approach. Helix uses exome sequencing, so it actually does read off the letters of your DNA. Your **exome** is the part of your DNA that is made into proteins. These are the parts that exit the nucleus, the **exons** you hopefully remember from the section titled "Transcription and RNA." That means the exome does *not* include most of the regulatory sequences that help to instruct the cell when and whether to transcribe the exons. (Helix calls their technique Exome+ because they sequence the exome *plus* a little bit of the regulatory sequences.)

Genotyping and exome sequencing are both shortcuts that allow companies to sell DNA analysis at consumer-level prices (a few hundred dollars, or less, depending on the company and the type of analysis they're selling). Full sequencing would cost more than a thousand dollars.

UNDERSTANDING YOUR DISEASE RISK

Genes Aren't Everything

With results from a personal genomics service, you can learn about your risk for certain diseases. Sounds pretty straightforward, right?

But there's a lot of interpretation that goes on between the time some of your DNA hybridizes to a chip in a genomics lab, and the time you sit there looking at your computer screen, thinking, "Huh, looks like I have a higher than usual risk for celiac disease."

First is the analysis that the lab does. Their system doesn't test every single base of your DNA, whether it's a genotyping or an exome sequencing service. Even the ones they do test can have errors. Often they adjust for this by testing each area of the genome more than once.

For example, with chip genotyping, there will be multiple spots on the chip looking for the same variant. And the computer that reads the results will compare the red, yellow, or green dot from each spot to other people's red, yellow, or green dots to make sure it's understanding correctly what color each spot is lighting up.

After that comes the job of matching up specific SNPs (single nucleotide polymorphisms, pronounced "snips") to what they mean in real life.

OBESITY: GENETIC, BUT MORE THAN JUST GENES

For example, one gene called FTO is known to be related to body weight and hunger. But the different genome services give slightly different explanations of what your variants in this gene happen to mean. EmbodyDNA describes a SNP called rs9939609 as indicating whether you benefit from exercise when it comes to losing weight. Another service, Habit, uses the same SNP to say whether you have an increased risk of obesity.

It turns out there are many variants in the FTO gene that could impact body weight. Scientists haven't quite figured out what each of them means, and they still don't know all of the ways that FTO interacts with other genes in your body.

FTO variants are common in personal genomics test results because the FTO gene is one that has been studied. We probably have tons of other genes that affect body weight, but not all of them show up in genomic tests.

And then another thing to add on top of that: genes aren't everything! If you have genes that should make you skinny, but you work a stressful job that requires night shifts and your coworkers are always bringing in doughnuts, you're likely to gain weight. On the other hand, if you decide you'd like to be healthier, and you go for a run every day before work and you try your best to say no to the doughnuts, your body weight will decrease. Even so, as you get older, your weight is likely to creep up again. Your genes just don't have enough information to tell you everything that's going on.

Obesity is a complex trait, and we understand that it's partly genetic but also depends a lot on environmental and lifestyle factors.

Many diseases are the same way: your heart disease and diabetes risks depend on how you eat, for example. Lactose intolerance is clearly a genetic trait, but if you have just the right mix of gut bacteria, you might never get severe symptoms no matter what genes you have.

Personal versus Medical Tests

Personal genomics tests are designed to give interesting results to most of the population. Meanwhile, medical genetics tests are more expensive and test for fewer things, but they are geared toward diagnosing a disease. If you want to know whether you carry a cystic fibrosis gene, for example, a personal genomics service will look for a few common variants. But a test ordered through your obstetrician will test for many more variants, including rare ones. If you have a serious medical concern about yourself or your future children, talk to your doctor or a genetic counselor.

SERIOUS DISEASES

Would you want to know if you have a serious medical condition, like an increased risk of developing Alzheimer's disease early in life? Some people want to know as much as they can, while others would rather wait and see.

It can be hard to wrap your mind around the idea that you have a deadly condition, or a greater than average chance of developing one. Some results are devastating: for example, if you find out that you have a gene for Huntington's disease, you have to face that fact that you will almost certainly develop uncontrollable motions, cognitive decline, and death—often in middle age.

There's nothing you can do to prevent this disease if you have the gene. That said, you might want to prepare for it, perhaps by writing a will and planning for your children's future.

On the other hand, some disease risks are actionable. If you know you have the BRCA1 or BRCA2 mutations that lead to an increased risk of breast and ovarian cancer, you can make sure to get mammograms and other tests more often. Some people with severe variants in these genes choose to have their ovaries and/or breasts surgically removed, to reduce the risk of developing cancer.

MORE QUESTIONS

If you like the idea of finding out more, would you want to know everything there is to know about a baby? Finding out in your thirties that you have more slow than fast muscle fibers is a curiosity: "Oh, that's funny, because I was actually better at sprinting," you might say. But if you learned that at age twelve, or if your parents knew that about you as an infant (before you were even born, perhaps?) would that have affected what sports you chose to play?

For these reasons, it might be best *not* to find out the hidden genetic traits of a child. And along the same lines, maybe you don't want to know everything about yourself, either.

WHERE HUMANS CAME FROM

Africa, for Starters

Even before we were human, we were African.

There, somewhere between thirteen million years ago and four million years ago, a population of apelike creatures separated into the ancestors of humans and the ancestors of chimpanzees. We don't have fossils of these particular individuals, but we know from DNA that such an ancestor must have existed. Where we share 99.9 percent of our DNA with other humans, we share around 98.4 percent with chimpanzees. They are our closest relatives, even closer than gorillas.

But it's not like one day our ancestor had two babies, and one was a human and the other a chimp. Rather, our ancestors' journey to becoming human was a long and gradual one. It took at least four million years, after all. (Chimps' ancestors must have taken a similar path, but chances are you and I are more interested in the human branch of this family tree.)

The oldest fossils that are arguably human were found in eastern Africa. There is a seven-million-year-old specimen from Chad, a six-million-year-old find from Kenya, and a few from four to five million years ago that were found in Ethiopia.

Those Ethiopian fossils were placed in a genus called *Ardepithecus*. (As the human species *Homo sapiens*, our genus is *Homo*.) Around four million years ago, individuals from the *Australopithecus* genus walked around eastern Africa. Both of these groups are extinct today; *Australopithecus* died out about two million years ago.

About 2.8 million years ago, our ancestors looked and acted enough like humans that when anthropologists find their fossils, they put them in the genus *Homo* along with us. First there was *Homo habilis*, with a chimpanzee-sized brain and the ability to use tools. Next come our larger-brained relatives *Homo ergaster* and *Homo erectus*.

The exact age of each species, and our understanding of what each one could do and where they lived, changes as new fossils are found. Some of the details are the subject of debate among scientists: for example, whether a new fossil should be its own species or if it belongs to one that's already been discovered. The relationships between the different species are also subject to change as we learn more.

One thing is clear: there isn't a single evolutionary path from our ancient ancestors to modern-day *Homo sapiens*. Our ancestors' family tree branched like any family tree. Some of those branches, like *Homo erectus*, lived in parallel to other branches but eventually died off without leaving children. There may also have been interbreeding between some of the branches of the family tree, making it hard for today's scientists to determine exactly who was who.

HOW HUMANS LEFT AFRICA

By about fourteen thousand years ago, there were humans all over the globe. No longer was Africa our only home; humans had spread to Europe, Asia, Australia, tons of Pacific islands, and both North and South America. So how did we get to all of these places?

First our ancestors from eastern Africa migrated to other parts of Africa too. Africa isn't just the home of one single lineage of humans; it contains many different branches of the human family tree.

Next, around sixty to seventy million years ago, a group of humans left Africa. (There were probably other migrations out of Africa before this, but as far as we know they didn't stick around.) This group ended up populating areas that include India, Southeast Asia, and even Australia.

After that, another wave of migration went through the Middle East to Central Asia and Europe.

Another major migration happened around twenty thousand years ago, when people from Asia made their way to North America, around what is now Alaska. These people slowly moved south and east, and by about fourteen thousand years ago there were humans living at the southern tip of South America.

YOU MIGHT BE PART NEANDERTHAL

Somewhere around half a million years ago, a group of people called the Neanderthals lived in Europe. Their ancestors seem to have left Africa before ours did. Some of their fossils were discovered in the 1800s, including one in a cave in the Neander Valley in Germany.

Neanderthals are usually classified as a separate species from modern humans: we are *Homo sapiens*, they are *Homo neanderthalensis*. Some scientists prefer to think of Neanderthals as a subspecies: *Homo sapiens neanderthalensis*, so closely related to modern humans that we should be considered the same species.

At least four hundred Neanderthal specimens have been found (skeletons or parts of them) across Europe and the Middle East. They had brains at least as big as ours, housed in skulls that featured thick, bony eyebrow ridges. They also had thicker bones than we have and probably stronger muscles. They made tools, used fire, and were hunters. They may have made art and purposefully buried their dead.

Some Neanderthal remains have DNA that has survived well enough that their genome can be sequenced. DNA tends to break into pieces after cells die, so ancient DNA is particularly hard to sequence. Scientists from the Max Planck Institutes came up with a way of sequencing the fragments of DNA, ignoring the DNA of any bacteria that may have snuck in. (Bacteria are found in soil and pretty much everywhere, so they can easily end up in samples of ancient DNA.)

The result, announced in 2010 and refined in studies since then: Neanderthal DNA is similar to ours, and it looks like Neanderthals and *Homo sapiens* interbred at multiple points in our history. As a result, some of us can trace a few genes back to our Neanderthal relatives. If you are European, American Indian, or Asian, you may have about 2 percent Neanderthal DNA. If your ancestors are all from sub-Saharan Africa, you probably have none.

A branch of the Neanderthal family, the Denisovans, lived in an area ranging from Siberia to Southeast Asia. (They are named after the Denisova Cave in Siberia.) People with Melanesian or Aboriginal Australian ancestry likely carry a portion of Denisovan DNA, perhaps 3 to 5 percent.

HOW MUCH DNA DO YOU SHARE WITH A STRANGER?

A Lot, It Turns Out

There's a huge amount of human diversity across the world. Even if we're only looking at traits that are genetic (or mostly genetic), we can see many different features. People can be tall or short, light-skinned or dark-skinned, or anything in between. We can have straight hair or curly hair, in shades of blond, red, brown, or black, with a variety of textures. Some people are more prone to cancers or other diseases, while other populations' genetics protect them at least a little bit. We have different shapes of noses, lengths of fingers, textures of earwax. There are so many things that make each of us unique.

And yet—our DNA is, at minimum, 99 percent identical.

Imagine meeting your counterpart from halfway around the world who is different from you in every genetic way you could think of. (Assume for the moment that they have the same chromosomes as you—so if you're XX and don't have any trisomies, they are the same way.) If you both get your genomes sequenced and compare notes, you'll find you aren't that different at all. You and that totally completely different person are *still* going to have more than 99 percent of your DNA in common.

WHAT'S IN OUR DNA?

The human genome, you might remember, is about three billion base pairs long. In other words, that's the number of nucleotides on all twenty-three of our chromosomes put together. (How many *you* personally have depends on whether you have both an X and a Y chromosome, but either way this number is approximately correct.)

There are about twenty-two thousand genes, or sets of instructions, for making proteins, and about three thousand for making functional RNAs (like ribosomal RNA, tRNA, and others that do specialized jobs). That's a pretty small number, especially when you consider our body makes close to 100,000 proteins. It turns out a lot of our genes can be alternatively spliced: the same instructions, followed in a slightly different way, can make two or more different products.

Conserved

Mutations occur all the time, but not all areas of the genome are equally subject to change. A mutation that occurs in the middle of an important gene or regulatory region could result in the organism dying. Meanwhile, a mutation in a noncoding area, or in a gene that we could live without, is more likely to be passed on without any problems. That means that the more important a region of DNA is, the less likely it is to change over time. We call these regions or genes "conserved," because they are preserved while the DNA around them changes.

All of those instructions put together are called the protein-coding regions of DNA (because they contain the code for building proteins). And they only add up to 1.5 percent of our genome.

Even when you consider all the parts of DNA that don't make proteins or RNAs, but that are important for regulating gene expression, we're only looking at another 6 percent of the genome. These regions are the **conserved** noncoding elements, and we know they are important because they are similar between people and, often, between species.

Only a little bit of our DNA actually makes a difference to our phenotype. Compared to other species, we have a "low gene density," meaning that any given stretch of DNA contains occasional genes in a sea of noncoding sequences.

LARGE PARTS OF OUR CHROMOSOMES DON'T CODE FOR ANYTHING

It's tempting to call the rest junk, but much of it does have a function, albeit indirectly. In 2012, a project called ENCODE announced that 80 percent of the human genome has some kind of function. A few years later, another team of scientists said no, the functional stuff is less than 10 percent. Both teams were correct, but they were each using different definitions of "functional." Most scientists agree that the smaller number is a better way of describing how much of our genome has a function that relates to gene expression.

Let's take a quick look at that 80 percent. The ENCODE scientists got that number by including segments of DNA that are often methylated, which means they are turned off. However, that means that our cells are paying specific attention to those segments and adding the methyl groups—so they're not passive pieces of junk but sequences that, in a sense, we do something with. The scientists also counted sequences of DNA that are wrapped around histones (a *lot* of our DNA is wrapped around histones, just to keep it from getting tangled). As well, they counted sequences that look like they bind to proteins, even if we don't have any evidence that those proteins exist or that they do anything interesting after they bind.

So most of our DNA isn't *actually* functional. Among the extra stuff, we have sections of chromosomes that have gotten duplicated and reinserted. We also have repetitive sequences, just the same few letters over and over again. Nearly half of our genome is made of transposons, sequences that can actually remove themselves from our genome and reinsert themselves elsewhere. These sequences are probably left over from ancient viruses that infected us but turned out to be mostly harmless. Because they can copy themselves, we end up with multiple copies of their DNA.

One class of **retrotransposons**, or elements that can copy and paste themselves, is the long interspersed repeat sequences, or LINEs. These are each a few thousand base pairs long, and together they make up 21 percent of our genome. There are also short interspersed repeat sequences, or SINEs, that make up another 6 percent.

HOW PERSONAL DNA ANCESTRY SERVICES WORK

Searching for Your Forebears

Personal genomics services don't stop at describing your individual genes. Many say they can tell you where your ancestors come from and give you clues to your family history and ethnic heritage. There are three main techniques they can use: analyzing the Y chromosome, mitochondrial DNA, or your autosomal DNA. Many services use more than one.

Y CHROMOSOME ANCESTRY

If you are a man with typical male genetics—an X and a Y chromosome—you can pass an X chromosome to half your offspring (your daughters) and a Y chromosome to the other half (your sons). Your son's Y chromosome will be an exact copy of yours, just as yours was an exact copy of your father's.

This identical copying makes the Y chromosome different from autosomes and even the X chromosome. Those chromosomes occur in pairs, so they get plenty of chances to recombine with their partners. When you pass chromosome 8 on to your child, just to pick a random example, that child may actually get a mix of your chromosome 8 from your Irish mom and your other chromosome 8 from your African dad. (For a refresher on why and how this recombination occurs, refer to earlier sections.)

Since Y chromosomes pass unchanged from father to son, that means that if you have a Y chromosome, you know you got it from your father's father's father's father's father (and so on). That means your Y chromosome traces a lineage of specific people, not a reshuffled deck of cards from many sources.

Everybody on Earth who has a Y chromosome can trace it back to a man who lived between 100,000 to 200,000 years ago. That ancestral Y chromosome got a mutation, making two distinct versions of the Y chromosome moving forward. Each of those underwent its own mutations over the years. Ancestry testing DNA services can tell you which branch of that Y chromosome tree, or which **haplotype**, your Y chromosome belongs to.

One major disadvantage of testing your ancestry through the Y chromosome: only people who have a Y chromosome can do it. If you only have X chromosomes though, you could try asking a brother, father, or paternal grandfather to find out their haplogroup and share the result with you. The results will still describe your ancestry, since they tell you which group your father's father's father (etc.) is part of.

Here's an important caveat, though: your (or your father's) Y chromosome haplogroup only tells you part of your ancestry. Your mother, for example, probably has a totally different ancestry through *her* father's father's (etc.) father. For the most complete results, and a great discussion topic at your next family reunion, ask to compare haplotypes with your mother's father, and your mother's mother's father, and so on.

MITOCHONDRIAL ANCESTRY

Just as males pass a Y chromosome to their offspring, females also pass on a unique genome: their mitochondria. (For more on why and how, refer to the "Mitochondrial DNA" section.) The mitochondria are part of the egg cell, so mothers pass mitochondrial DNA to all of their children, regardless of sex. But only their daughters who produce egg cells and conceive children with them will be able to pass on those mitochondria to the next generation.

Haplotype versus Haplogroup

A haplotype is a collection of alleles that are shared by members of a haplogroup. For example, about 27 percent of male Australian Aborigines carry a certain collection of alleles on their Y chromosome. Together, those people and their ancestors make up a haplogroup known to scientists as S1a1a1.

That means that your mother gave you the mitochondrial DNA from her mother, which came from her mother, and so on. This gives us another recombination-free family tree, and you can have a mitochondrial haplotype too. (A haplotype means any collection of DNA variants that are passed on together, but when we talk about haplotypes in humans, we're usually referring to either the Y chromosome or the mitochondrial genome.)

AUTOSOMAL ANCESTRY

Some services analyze your ancestry based on your autosomal DNA (chromosomes 1 through 22). We all have these chromosomes, so this type of testing works for both males and females. It can turn up results from anywhere in your family history, and isn't limited to just your mother's or your father's heritage.

Autosomal testing works by looking for SNPs that you share with other people in the database. But instead of looking at one variant at a time, the analysis looks for groups of SNPs that seem to be inherited together. These groups aren't necessarily on a single chromosome, but can be a constellation of SNPs from all across the genome.

Mitochondrial "Eve" and Y Chromosome "Adam"

Since all humans are related, we can all trace back our Y chromosomes to a single ancestor. We can *also* trace back our mitochondrial DNA to a single ancestor. But these two ancestors are not the same person—one must be male and the other female, of course. They also don't necessarily come from the same part of the world or the same point in time. After all, the common ancestor of all human mitochondrial DNA, sometimes nicknamed "mitochondrial Eve," wasn't the only human being on Earth! She probably had sisters and cousins and distant relatives who just didn't happen to have a line of descendants that stretches to the present day. Geneticist Allan Wilson suggested a more accurate nickname: the lucky mother. Her male counterpart, who she probably never met, could then be called the lucky father.

RACE, ETHNICITY, AND ANCESTRY

Individuals and Larger Groups

Since DNA helps scientists figure out how people and other living things are related, it makes sense that a DNA test could tell you where you fit in to that big old family tree. Indeed, a bunch of DNA testing services exist that say they can tell you who your ancestors are. But you'll want to proceed with caution. While they can tell you a lot about who you *might* be related to, they also have some drawbacks that you should be aware of before you start.

ETHNICITY IS HARD TO PIN DOWN

The first thing you have to realize when you're trying to find out where you're "really" from is that race and ethnicity aren't written in stone in your DNA. Race and ethnicity lie at a complex intersection of biology, history, and culture. Our genes often don't match up to what our cultural history tells us our ethnicity "should" be.

Let's start with ethnicity. We usually think of it as relating to national borders: perhaps family lore holds that you are German, or Ethiopian, or Japanese, or native Hawaiian. But what does it really mean to be "German," especially since that's a country whose borders have shifted several times in the past one hundred years? Even when borders are stable—take Japan, for example—when does a person become Japanese? When they have lived there for a few years? A generation? Five or ten generations?

Populations don't keep to themselves, either. Even if a group of people has lived in the same area for a long time, they are rarely totally isolated. Groups of people routinely intermix, either by happily marrying people in nearby communities, or by tragic means like enslavement and rape. Either way, this complicates our idea of a family tree with distinct branches. It turns out that rather than a tree, human heritage looks more like a net, with relationships crisscrossing everywhere.

RACE DOESN'T REFLECT FUNDAMENTAL DIFFERENCES BETWEEN PEOPLE

Your DNA doesn't reflect geography, only relationships. That's one major reason why DNA tests will never totally satisfy your curiosity about where your recent ancestors were born. However, with enough data, we can figure out whether your ancestors were closely related to other people who think of themselves as, say, Japanese. Consider your DNA one of many clues to figuring out your family history.

Race is even more nebulous. This may sound confusing if you grew up thinking of races as totally separate groups of people, each with their own lineage. It's tempting to think that, because the categories we call races are often defined by some visually obvious traits like skin color, hair texture, and body shape. But those traits actually *don't* hint at bigger differences. People of vastly different "races" have very similar genetics otherwise.

Evolutionary biologist Richard Lewontin calculated that of all the variation in the human genome, most of it occurs *within* groups: differences between you and your neighbors. Only about 15 percent of the variation in human genetics is *between* groups of people. Some amount of skin color variation is in that 15 percent, along with gene variants that evolved relatively recently, like the ability to digest lactose as an adult. But most of the differences that make us unique, from intelligence to blood type, are present in people of every major population group.

Is Race Real?

Race isn't a fundamental underpinning of our biology. But that doesn't mean it's a figment of our imagination, either. The features that we think of as defining races, like skin color, really do exist; they just don't indicate anything special. Skin color is only skin deep. Meanwhile, race is *very* real as a social construct. Whether people are mistreated or welcomed in a community often depends on the race that they appear to be.

WHAT DNA CAN TELL YOU ABOUT YOUR ANCESTORS

Here's what DNA tests *can* do: they can tell you who you're most closely related to, out of a large group of people who have also had the same kinds of tests done. Now that databases are keeping track of millions of people at a time, we can definitely notice some patterns in who is related to whom.

So even though you can't always perfectly determine your race or ethnicity, you *can* often make a good guess about who you're related to. There's a good chance that a lot of people living in Japan, for example, share genetic variations with each other that they can trace back to some of the first people who settled in Japan. If you grow up in the United States, not knowing your ancestry, and then get a DNA test, you may find that your genes match those people in Japan more closely than people in other parts of the world. You can then make a good guess (but not a guarantee) that at least some of your ancestors came from Japan.

These testing services are only as good as their database, though. First, some personal genome services have bigger databases than others. The bigger the database, the more accurately you can determine who you are related to. The second issue with databases is that so far, most of them include a lot of individuals with European ancestry and typically not as many with other types of ancestry, such as Asian or African. As a result, they will give the most specific results for Europeans: perhaps your DNA matches pretty well with people from a certain area in Ireland. However, if that same database only has a handful of data points in Africa, it won't be able to pinpoint your African heritage the same way. You might get a match that can only be localized to, say, the eastern half of the continent.

WHAT ABOUT DOG BREEDS?

Let's contrast the human family tree with that of another species that shows a lot of variation worldwide: dogs. Dog breeds are diverging branches of a family tree. They are exactly what human races or ethnicities *aren't*.

Let's look at what this means. Domestic dogs are all one species, whether they're tiny teacup Chihuahuas or monstrous, furry Akitas. They all descended from an ancestor that may have looked something like a wolf.

Thousands of years ago, perhaps around the time hunting and gathering fell out of fashion in some parts of the world, people began breeding their dogs with others who had the same characteristics. For example, dogs that were good at herding would be bred with other dogs that were also good at herding. That was the beginning of the separation that later was refined into the dog breeds we know today.

Unlike with humans, dog breeds really were kept separate from each other over a long period of time. A recent genetic analysis of 161 breeds (about half of the known dog breeds) found that the phylogeny, or family tree, of modern dogs has twenty-three large branches called **clades**. One clade includes spaniels and retrievers, for example. Another includes bulldogs and boxers. Yet another is full of terriers.

Dogs in each clade mated with other dogs within the same clade, for the most part. This makes sense: if you want terrier puppies, you breed two terriers together. Within the last two hundred years or so, people started getting even more specific about the name and identity of each breed, so each branch of the family tree has divided into many smaller branches that don't mix with each other. Scientists were able to trace back this whole history and were also able to tell when certain breeds were crossed into others deliberately. For example, there is a group of dogs bred for fighting that come from crosses of terriers and bulldogs in Ireland in the mid-1800s.

This only describes purebred dogs, of course. Plenty of mutts don't fit neatly into any breed category because they may be the descendants of mutts going back generations.

It's possible to get DNA tests to trace your dog's ancestry, and these work best when your dog has recent ancestors (parents or grandparents) that come from one of these well-defined breeds. But the results can be inconsistent, and again they're only as good as the testing company's comparison database.

RELATING TO YOUR RELATIVES

Just How Close Are You?

Now that we understand how humans relate to one another on the great big family tree of life, let's zoom in and look at *your* family, in particular.

First, let's try to put a number on how closely related you are to each person in your family. You probably already understood this even before you started learning about genetics. For example, if your mother is from Germany and your father is from China, you might tell someone that you're half German and half Chinese. Or if you know that one of your grandparents was from Greece, you might say you are a quarter Greek.

These numbers make sense from a genetic standpoint too. You get half your chromosomes from your mother and half from your father, so you really do share 50 percent of your DNA with each parent.

Let's look back another generation. Your mother shares half of her DNA with each of *her* parents, so you probably share about 25 percent with her mother, who is also your grandmother. Makes sense, right?

But you aren't guaranteed to be *exactly* 25 percent identical to your grandma. That's just an average. Remember, when your mother made the egg cell that became you, her cells went through meiosis and randomly picked one chromosome from each pair to pass on to you. Of the twenty-three chromosomes she gave you, odds are that eleven or twelve of them were from her mother, and the rest were from her father.

Flip a coin twenty-three times, and you'll probably get eleven or twelve heads. But it's possible, if rare, to flip a coin twenty-three times and get twenty-three heads. And in meiosis, every now and then the resulting egg cell will have twenty-three chromosomes from one parent and none from the other. That means you could share anywhere between zero and 50 percent of your DNA with each grandparent. Most of the time it will be close to 25 percent from each, but the split between the two grandparents of a pair could instead be 24 and 26 percent, or 31 and 19 percent, or 1 and 49 percent.

By the same token, you share *about* 50 percent of your DNA with a full sibling, but that could vary as well. The egg cell that made me and the egg cell that made my brother could have had the same selection of chromosomes (we might look like twins), or we may have lucked out and gotten the opposite chromosome of each pair (my maternal chromosome 1 might come from our maternal grandmother, while his maternal chromosome 1 might come from Grandpa).

How Related Are We?

Here's how related you are to your relatives, on average:

- 100 percent with an identical twin
- 50 percent with your parent or your child
- ~50 percent with a full sibling
- ~25 percent with a grandparent, grandchild, aunt, uncle, or half sibling
- ~12.5 percent ($\frac{1}{8}$) with a great-grandparent, a great-grandchild, a great-aunt or great-uncle, or a first cousin
- ~6.25 percent ($\frac{1}{16}$) with a great-great-grandparent or great-great-grandchild
- ~3.13 percent ($\frac{1}{32}$) with a great-great-great-grandparent, a great-great-great-grandchild, or a second cousin
- ~0 percent with a stranger

We are each 50 percent Mom and 50 percent Dad, but in each case my 50 percent might be different than my brother's.

The numbers we're talking about here are measuring **identity by descent**. This doesn't measure how much of your DNA might be identical to another person's (we'll get to that in a minute) because, don't forget, your DNA is 99 percent identical to a stranger's and 98 percent to a chimpanzee. Rather, identity by descent reflects how much of your DNA you got *from the same recent source* as the person you're comparing yourself to.

IDENTITY BY STATE

Because we're all so similar, there's no way a DNA test can give you the numbers mentioned previously. For example, maybe all four of your grandparents had the same allele for lactose intolerance, which is totally possible because a good chunk of the human population has the same allele they do. There's no genetic test that can tell the difference between alleles that are genetically identical.

That's why personal genome tests that use genotyping, like 23andMe, calculate relatedness differently than we did earlier. They sample a bunch of sites across your genome, about 600,000 in the most recent version of the company's chip. These are from sites that are known to vary between people, so they're skipping the places that are identical in just about everybody.

Even so, we are all fairly closely related. Chances are you would match a perfect stranger at about 70 percent of those single nucleotide variant sites. An identical twin, on the other hand, would match you at 100 percent.

The results for each SNP take the form of which nucleotides you have on your two chromosomes. For example, a SNP locus called rs4988235 comes in two common versions, one with a C in the genetic code at that point, and the other with a T. If you have the C allele on both chromosomes, your genotype is CC, and you're probably lactose intolerant. Your parent might be CT (50 percent identical to you, at that locus) or CC (100 percent identical). To find out how related you are to a person, the genetic testing service compares you to the other person across all of the SNPs they checked.

How Much DNA Do We Share?

Here's how many SNPs you're likely to share with different relatives, assuming a selection where you share 70 percent with a stranger:

- 100 percent with an identical twin
- 85 percent with a parent or child
- ~85 percent with a full sibling
- ~77.5 percent with a grandparent, grandchild, aunt, uncle, or half sibling
- ~73.75 percent with a great-grandparent, a great-grandchild, a great-aunt or great-uncle, or a first cousin
- ~71.9 percent with a great-great-grandparent or great-great-grandchild
- ~70.9 percent with a great-great-great-grandparent, a great-great-great-grandchild, or a second cousin
- ~70 percent with a stranger

A service that uses genotype results to look for relatives will compare each other person's results to yours, using those 70 and 100 percent numbers as a reference (or whatever the numbers are with their specific selection of SNPs). You're 50 percent related to each parent, so your identity by state score should be about 85 percent in this case: halfway between 70 percent (what you share with a stranger) and 100 percent (which would be somebody identical to you). If the service finds another person who matches 85 percent of your SNPs, it guesses that that person is either your parent, your child, or your sibling. And somebody who shares 92.5 percent of your SNPs is probably 25 percent related to you by descent—so they would be either a grandparent or a half sibling.

ORDER MATTERS

We know where each SNP occurs on the chromosome, so we can sometimes use those positions to understand where each of your SNPs came from.

For example, if you're looking at a comparison between you and your parent, one of each of your chromosomes will entirely match theirs. But if you're looking at a sibling, what you share with that person will be a patchwork of DNA from both chromosomes of each pair.

Why a patchwork? Remember what happens during meiosis: not only are chromosomes dealt out to one cell or another, but in the process the chromosomes swap pieces. Each chromosome that's passed down (say your mother's chromosome 1 that she got from her mother) will have bits of its mate (in this example, pieces of your mother's *father's* chromosome 1). Your sibling could get the same chromosome as you or a different one, but at the same time they are also getting a chromosome that went through a different set of swaps in recombination.

A FAMILY TREE FOR ALL LIFE

We're All Related

Not long ago we talked about family trees. Let's draw one again, but make it bigger. To avoid wasting paper, you can do this in your mind. (If not, you would need a *lot* of paper.)

Start with yourself. If you have siblings, then you know you're connected to them through your parents. That's easy, right?

Now draw a little more of the tree, so your first cousins show up. You and your favorite cousin share a pair of ancestors too: specifically, your grandparents. Keep going, and you'll see that you and your second cousins can name your great-grandparents as your most recent common ancestors. If you had access to perfect and complete genealogical records for everyone in the whole world—which you don't, nobody does—you would be able to draw the family tree big enough that it includes literally everyone in the world, and everyone they were descended from. See why I told you not to get out your good paper?

It wouldn't include everyone who ever lived, though. Maybe you have a great-great-aunt who never had any children. She wouldn't be on your chart, even though she did exist. There are plenty more people, including whole branches of your family, who don't have any descendants that are alive today. There may even be whole towns or whole countries whose populations didn't make it to the present day.

ONE BIG FAMILY

You don't have to stop at humans, although by the time you're making a tree this big you probably aren't going to sketch in every individual creature. Instead of Koko the gorilla, you can just have a spot on the family tree to represent the entire subspecies of western lowland gorillas. They would be closely related—like siblings, you could say—to the subspecies of Cross River gorillas. Those groups together are the lowland gorillas, and they are related like cousins to the mountain gorillas.

Meanwhile, humans and chimpanzees also have a (more distant) cousin-like relationship, and then gorillas are related to us, but even more distantly. Other apes are cousins too, and then monkeys and lemurs are our very, very distant cousins. We have an even more distant relationship to other types of mammals and to other types of animals. Take this tree back far enough, and we'll see our very, very, *very* distant relationships to reptiles, and to fish, and to insects, and even to things that aren't animals at all. We even share a common ancestor with plants. That ancestor, our great-great-great-(insert a lot of "greats" here)-great-grandparent, was a single-celled organism. It probably looked like a blobby microscopic creature called a protist. Protists still exist today, and you can see them if you scoop up some pond scum and look at it under a microscope. No matter who you are, you can say you came from humble beginnings.

Protists, in turn, can trace their family tree back to even simpler single-celled creatures. The oldest forms of life we know of are bacteria and archaea. The archaea, whose name means "old ones," actually turned out to be a newer branch of the family tree than bacteria. But by the time scientists figured that out, people were too used to the name to change it.

FINDING RELATIVES

I see you out there, wearing down your pencil on sheets of paper taped together. I told you not to use paper, but fine, I see you are determined. But you're stuck. *How do I know which creatures are related to which?*, you say.

In the olden days, people would look at living things and judge them to be related based on how similar they looked. Dogs and wolves and coyotes look like the same sorts of creatures. Lions and housecats have a lot in common. The cat family must be related to the dog family somehow, because they are all four-legged furry creatures.

This system works okay most of the time, but pretty quickly you'll run into trouble. Do bats go with mammals, because they are furry? Or with birds, because they fly? You can examine different aspects of a bat and find that it's far more similar to rodents than to birds: bats give birth to live young and feed them with milk, for example. They have teeth, not beaks. And if you study how their skull and ear bones are put together, it's pretty clear that they're related to other mammals.

So even before DNA analysis came on the scene, scientists knew to look for family relationships among plants, animals, and anything else they studied. Sometimes they would argue over exactly which group something belonged in, but even before the discovery of DNA there was no doubt that living things are related to each other in family trees that branch just like ours.

DNA AS DATA

Once we knew that the sequence of letters in DNA meant something, scientists started to get curious about exactly what letters were in the DNA of each living thing. At first it took months of work in the lab to come up with even a short sequence of DNA, but over time, scientists have gotten better and better at reading DNA sequences. There's even a word for this process: **sequencing**.

Sequencing the Human Genome

In 1990, a group of scientists at universities and institutions across the United States launched a massive project aiming to read and catalog the entire sequence of human DNA. They figured it would take fifteen years but managed to finish early, by 2003, which was the fifty-year anniversary of Watson and Crick's discovery of DNA's structure. Sequencing technology has gotten drastically cheaper and faster over time. Today, there are labs that can sequence a human genome in just a few hours, for a cost of around $1,500.

Now that we can sequence genes, or even whole genomes, we can compare them. Remember how you have a few mutations that are all your own, that your parents didn't have? Those make you a little different from your siblings too. There will be a few differences between you and your brother and more between you and your cousin. If you could compare the complete DNA sequences of everybody in the world, you could use those tiny differences to reconstruct that entire big family tree you were trying to draw out on paper. All you need is enough computer power to do it.

And this is what many scientists have been doing ever since they've been able to sequence DNA. Even back when it was too difficult and expensive to sequence an entire genome, they could sequence just one gene and compare that same gene between species. The more similar the sequences, the more closely related the species they came from.

Back when scientists were drawing family trees for living things based on their appearance and structure, they called that field of study phylogeny. Now that we can do the same thing with DNA, it's called **phylogenetics**. Amazingly, the family trees created by DNA and by old-fashioned observation pretty much matched. But in a few cases, DNA showed that we didn't have the family tree drawn quite right.

That's how we learned that bacteria and archaea are different (even though they look similar under a microscope) and how we learned that the archaea are actually a younger branch. It also turns out that elephants are related to aardvarks and a few other odd-looking smaller creatures, like elephant shrews (so called because they have long noses). On the plant side of things, a group of geneticists meets every few years to check in with DNA data and shuffle plant species around from family to family, to be sure that botanists are working with the most up-to-date family trees.

By sequencing DNA from different people and ethnic groups across the planet, we can learn a little more about how we are all related to each other.

EVOLUTION

How It Happens

"Nothing in biology makes sense except in the light of evolution," biologist Theodosius Dobzhansky once wrote. Evolution explains how species are related to each other, and it also explains why each living thing has particular features and quirks.

Here are some examples of evolution: humans today have bigger brains than our ancestors who existed millions of years ago. The finches on the Galápagos Islands today have different beak sizes than the finches that lived on those islands just a few hundred years back. And—to bring this a bit closer to home—the bacteria in your throat look different after you've taken a weak dose of antibiotics (oops, should have taken all the doses as prescribed) than they did at the start of your illness.

Here's what all of these cases have in common: they all represent a *population* whose gene pool has changed over time. In the human example, more of us have the genes that allow for bigger brains than that ancient population had in the past. In your throat, perhaps only a few individual bacteria had antibiotic resistance genes at the start. But after applying a selection pressure for a little while—weak doses of antibiotics—you have shifted that balance so that far more of the bacteria have those resistance genes.

This is what evolution really means: it's the change in allele frequencies in a population over time. Forget any images you might have of individual humans or bacteria slowly changing. Evolution doesn't happen to individuals. It's a property of multigenerational groups.

Let's take a look at a few things that can cause evolution to happen.

MUTATION

A mutation occurs when a new allele is added to the gene pool. For example, there was probably once a time when all humans were lactose intolerant; none of us had the ability to keep producing the lactase enzyme after childhood. Then somebody was born who had a mutation in the lactase gene. As soon as that mutation appeared, the population's gene pool had automatically changed: there was now one more allele than there had been before.

Mutations can become more common over time. Sometimes that's just a matter of chance; maybe the gene doesn't make much of a difference to survival, but the people who have it just happen to have a few more children. Other times, the mutation becomes more common for a specific reason, including some of the other items on this list.

MIGRATION

Evolution happens to populations, so it's significant if individuals move into or out of a population. If there's an island where all the birds have large beaks, and then a new group of the same species birds moves in, and they have small beaks, the gene pool of birds on that island has changed. Migration doesn't change features on a species-wide level, but it can influence what happens in that one population.

SHRINKING POPULATION

Populations can grow and shrink in size over time. For example, a population suddenly shrinks when there is an epidemic of a deadly disease. The people who were left in Europe after the Black Death killed 30 to 60 percent of the continent's population may have carried different alleles than the larger group of people who were present before the pandemic. Or take a look at any endangered species: by the time you get down to just a few hundred or a few dozen individuals, those individuals only represent a small sample of the original, larger gene pool. Even if they reproduce enough to bring the population back to its original numbers, the restored population won't have anywhere near the same diversity of alleles.

NONRANDOM MATING

Allele frequencies can also change if individuals don't mate at random. For example, if female peacocks prefer males with long tails, the genes related to long tails are likely to become more common in the population.

NATURAL SELECTION

All of the forms of evolution we mentioned previously are random. They happen, and they result in a changed population, but they don't reflect any direction or intention. They just happen because they happen, and that's that.

Natural selection, on the other hand, is what Charles Darwin was talking about when he first wrote about the concept of evolution. His book *On the Origin of Species by Means of Natural Selection*, published in 1859, described not just evolution but also **adaptation**, the phenomenon whereby living things become more suited to their environment. Darwin had no idea that genes were a thing and would never have dreamed that there might be a molecule called DNA at the heart of his theory. But his basic ideas made so much sense that they're still considered to be roughly correct today. Here's how he described the process of natural selection:

First, there is variation within a population. Darwin was thinking in terms of phenotypes; so, for example, a population of birds might include individuals with larger and smaller beaks.

Next, those variations can be inherited. Darwin didn't know about Mendel's work, so he wasn't familiar with the idea of a gene. But he had a sense that birds with large beaks would tend to have offspring with large beaks.

We know now that genes are responsible for those variations. If scientists are studying evolution or population genetics today, they don't settle for just observing characteristics of the creatures they're studying. Instead, they can also test DNA for characteristics they're looking for. Plant breeders, for example, use DNA tests to decide which individuals to breed together to make offspring with the traits they're looking for.

Finally, for natural selection to occur, the population has to be in a situation where not everyone survives to leave offspring. Think about a pond full of frogs. A pair of bullfrogs could easily produce over ten thousand eggs each year, and each parent lives an average of about eight years. Out of all the eggs they produce, only *two* need to survive to adulthood to replace their parents. If the population of bullfrogs in the pond stays the same from year to year, that means that many thousands of each frog's offspring will die.

Those numbers set up some stiff competition among the baby frogs. If one tadpole doesn't swim quite as well as the rest, it's more likely to get eaten by a fish than its many brothers and sisters. The two frogs that survive are likely to be fast swimmers, good at hiding, and also good at obtaining food and other resources to take care of themselves. Some of their characteristics are likely to be genetic. There are surely genes related to swimming ability, for example. As the slower tadpoles get eaten, the alleles related to fast swimming become more common in the population. Over time, the population of frogs will have more and more of the faster-swimming allele.

SPECIATION

So populations change over time, both for random reasons and because of natural selection. But that doesn't explain, by itself, why we have so many species. Why are there humans *and* chimpanzees in this world? Once upon a time we and chimpanzees had the same ancestors. What made our family tree branch?

Two populations can form where there was originally one. All it takes is millennia of separation. Perhaps an island becomes separated from a continent, and now the creatures on the mainland can no longer reach the island. The populations can't interbreed anymore, and so they evolve separately in their own ways. This explains why islands that have been isolated for a very long time tend to have species found nowhere else. Take the lemurs of Madagascar, for example, or the unique marsupials found in Australia.

Populations can also be separated by invisible barriers. Perhaps a few individuals in a population get their internal clocks mixed up, and they mate in the fall while their relatives mate in the spring. Pretty soon you'll have two separate populations that don't mate with each other, but live in the same place undergoing the same pressures.

DNA REPAIR

Fixing Mistakes

A job as complicated as copying DNA isn't going to go without a hitch. The enzyme that adds new nucleotides to a DNA strand tends to make a mistake every 100,000 nucleotides or so. Then once all that DNA is sitting around in a cell, stuff can happen to it. Free radicals can bump into it, or ultraviolet light from the sun can damage it. Radiation from hazardous materials or even the small doses in x-rays can cause damage to DNA as well.

Preventing DNA damage is important. That's why we wear sunscreen, avoid smoking, and trust our doctors not to order any more imaging tests than we really need. But when that damage inevitably happens, it's a good thing that our body knows how to fix it.

FIXING ERRORS IN DNA COPYING

DNA polymerase isn't perfect; it screws up one of every 100,000 nucleotides it adds. But our cells' machinery fixes most of those errors, taking the number of mistakes down to just one in a billion. That means each cell only has about six mistakes total, when you count both copies of our DNA. Not bad!

One process, **DNA proofreading**, happens during replication. When the DNA polymerase inserts the wrong nucleotide, it doesn't match up perfectly to the base on the other side of the strand. As a result, the new, wrong nucleotide's 3' hydroxyl group—the place where the next nucleotide will attach—doesn't fit into the right place on the DNA polymerase. The enzyme can't move forward until it removes the new, wrong nucleotide.

Occasionally, the DNA polymerase can move past an incorrect nucleotide. But that leaves a bump in the DNA, a wonky-shaped place where the nucleotides don't match up correctly. (Remember, each nucleotide only has one partner that enables it to fit properly in the double helix.) The enzymes that do this job, called **mismatch repair**, can spot the mistake because they know which is the old strand and which is the new one. That's thanks to methylation, one of the modifications discussed in the "Epigenetics" section. The new strand isn't methylated until after mismatch repair, so the enzymes can recognize the unmethylated strand as the new one.

These repair mechanisms are aimed at single nucleotide mutations, where a base pair is clearly mismatched. Mismatch repair can sometimes also catch a tricky mistake called a trinucleotide repeat. In these, a sequence like "CAGCAGCAGCAG" can cause trouble. If the two strands of DNA are separated, the trinucleotide sequence can fold and bind to itself, making a hairpin shape that hangs off the main DNA strand. If it's not caught, the next time this stretch of DNA is copied, the extra-long strand might be copied correctly, resulting in a sequence that is too long on *both* strands. Over time, this repeated sequence can accumulate more and more repeats. Some genetic diseases, including fragile-X syndrome and myotonic dystrophy, result from repeats like this that get out of hand.

FIXING DAMAGE TO DNA

Damage is easier to spot than errors, since the result is often a stretch of DNA that is not shaped or does not act like a normal double helix.

For example, pyrimidine bases (T or C) that are next to each other on one strand of the DNA can end up covalently bonded to one another after they have been exposed to ultraviolet light. That's obviously a problem, because DNA's bases aren't supposed to be attached to each other side to side. Some organisms, including bacteria, can spot these pairs and separate them. But the human approach is to just cut out both of the mangled bases and replace them.

In this and other cases where only one strand is damaged, enzymes can remove the bases that don't belong. When a single base is damaged, they can remove just that one. When multiple bases are involved, they often remove at least a dozen bases on the affected side of the strand.

Amazingly, our cells can even repair DNA when both strands have been damaged or broken. Because we have two copies of almost every chromosome (sorry, XY folks), double-stranded damage can often be fixed by matching up the broken ends to their homologous partners. This is almost exactly the same process that happens during meiosis, when pairs of chromosomes undergo recombination, or crossing over. In this case, each strand finds its partner on the homologous chromosome, and enzymes repair the DNA as the strands separate.

WHEN DAMAGE CAN'T BE FIXED

A cell with damaged DNA is a liability to our health. Think about what happens when we get too much sun over the course of our lives, or if we are exposed to carcinogenic chemicals or radiation. Some cells will be so badly damaged that they must die; others can end up causing cancer.

Cells are programmed not to replicate their DNA unless it is intact and undamaged. After that comes another point where the cell will not divide unless the DNA was properly replicated. Finally, there is another checkpoint during mitosis, at metaphase, where the spindle won't pull the chromosomes to each side of the cell unless each chromatid is appropriately attached to a microtubule.

If a damaged cell cannot be repaired, it self-destructs. This process is called **apoptosis**, and it results in small fragments of cells that can be gobbled up by our white blood cells (which already patrol our body looking for bacteria and other invaders to eat).

The real problem comes when a cell is damaged beyond repair but apoptosis does not occur. This can happen if a person has a mutation in one of the enzymes responsible for making apoptosis happen. For example, variations in the "tumor suppressor" gene that makes the p53 protein can cause a syndrome called Li-Fraumeni, where affected people have a 50 percent chance of developing some kind of cancer by the time they turn thirty, and a 90 percent chance by the time they are seventy.

Even if most of your cells have perfectly intact tumor suppressor genes, cancer cells can get a foothold if they end up with damage to their tumor suppressor genes. Cancer cells often have multiple mutations that help them keep growing when they shouldn't. These typically include mutations in one or more tumor suppressor genes.

CANCER GENETICS

Why Some People Are More Susceptible

DNA damage can lead to cancer, but our bodies have multiple safeguards to try to ensure it doesn't. As you read in an earlier section about DNA damage, cells can pause their division or even self-destruct if damage occurs and they are unable to fix it. But some people are more susceptible to cancer than others, and sometimes that relates to a mutation that makes them more susceptible.

When somebody says they have a family history of cancer, that means they probably have a gene that's been passed from generation to generation, with a detrimental mutation in some part of the safeguarding machinery. If that gene's purpose is to stop cancerous cells from dividing, an allele with a mutation is like a security guard with a tendency to fall asleep on the job.

Cancer is not a single disease. It's a collection of different diseases that have something in common: cell proliferation. That means cells divide when they shouldn't, leading to tumors that grow too large or that rob other cells of their nutrients. Cancers can arise for different reasons and in different types of cells. It's unlikely that there will ever be a single cure for cancer, but scientists are currently researching treatments, prevention, and potential cures for each of the hundreds of types of cancer people get each year. The most common kinds in the US are breast, lung, colorectal, and prostate cancer.

It usually takes more than one mutation for cancer to develop. Even in people with serious cancer-causing mutations, it can take years for the cancer to develop, and many people never get it at all. Most of that person's cells are fine, but one day one of those cells gets another mutation in just the right spot, and then it's able to grow out of control.

Cells with mutations that allow them to grow out of control will proliferate, and if one of those cells acquires another mutation with a similar effect, it could grow even faster. This is how mutations seem to accumulate. If one of those early mutations results in sloppier DNA copying or a loss of cell checkpoints, mutations can become more common and the cell won't be able to stop them.

SOME OF THE GENES INVOLVED IN CANCER SUSCEPTIBILITY

Oncogenes start as proto-oncogenes, normal genes that give instructions that tell a cell it's okay to divide. But with certain kinds of mutations, proto-oncogenes become oncogenes, encouraging cells to divide more often than they should. These genes can include growth factors, or the receptors for growth factors.

Tumor suppressor genes stop abnormal cells from dividing or surviving. A cancer-causing mutation in one of these genes results in the guard failing to do its job, so abnormal cells keep on dividing when they shouldn't. One well-known tumor suppressor gene is p53, which normally helps to initiate apoptosis.

Cell cycle genes are involved in telling cells when it's time to divide. Genes called cyclins and cyclin-dependent kinases help the cell to keep track of where it is in the cycle by turning on genes that are necessary for each step of the process. When these are mutated, they can result in the cell dividing even when it's not time.

DNA repair genes can also cause cell growth and cancer if they are mutated. Normally, a cell with damaged DNA should not divide, but if the repair proteins are broken and other genes are also mutated, the cell might go ahead and divide anyway. This kind of mutation can make further mutations more likely.

Telomerase is not normally present in most of our cells, so the ends of chromosomes, or telomeres, get shorter with each cell division. But cancer cells often have active telomerase, which probably helps them keep dividing past their normal lifespan.

Vascularization genes help tumors develop a blood supply to feed them nutrients. With an ample blood supply, the tumor can thrive and grow.

TUMOR SEQUENCING

Now that genetic tests and genome sequencing are relatively affordable, an oncologist can test the genome of a tumor—which, thanks to all those mutations, is different from the genome of the rest of your cells.

There are now targeted therapies available that can reverse the effects of certain specific mutations. For example, some non–small cell lung cancers have a mutation in a gene called EGFR, which is a receptor for a growth factor. The mutation means that the cell acts as if it's constantly receiving growth factor signals, even though the signals aren't actually there. If a tumor has this type of mutation in EGFR, the patient can take a targeted therapy like cetuximab, which blocks the receptor and turns off the growth signal. These therapies are expensive and may have serious side effects, but they are sometimes more effective than traditional chemotherapy.

PHARMACOGENOMICS

How Genes and Drugs Interact

Genes don't just affect how our body looks; they can also influence what our body does. That includes how our body uses medicines and other drugs.

What Is a Drug?

Drugs include any substance that has a specific effect on our body. Think of how we take caffeine (by drinking coffee or energy drinks) to keep ourselves awake. Or alcohol, which we hope relaxes us and/or makes us do stupid things. The category of drugs includes illegal drugs too, not to mention any of the thousands of medications we might end up taking under a doctor's orders.

After we get a drug into our body, enzymes work on that drug to break it down, or not. And the drug works by interacting with receptors or other proteins, enzymes, or parts of our cells. If you and I each have a different version of an enzyme that interacts with a drug, we might react in different ways to the same medication. Let's look at a few examples.

CAFFEINE METABOLISM

Caffeine gets broken down in our liver after we ingest it. Before a molecule of caffeine gets broken down, it rides through the bloodstream, occasionally stopping off at the brain to block adenosine in brain cells to keep us awake. But whenever it flows through the liver, it can be broken down by enzymes located there. Breaking down chemicals in our bloodstream is a big part of our liver's job, and it has tons of enzymes to help do this job.

One of those enzymes is called Cytochrome P450 1A2, abbreviated CYP1A2. There is a single nucleotide variant in this gene: a large chunk of the population has an A allele at a certain position, and other people have a C. (Other mutations are possible in this gene, of course, but these are two very common ones.)

Enzymes made with one of these alleles work at a slow and steady pace, letting plenty of caffeine circulate through our bloodstream in the meantime. The other variant does its job more quickly, so that cup of coffee you drank to keep yourself awake will be gone in no time. People who have either one or two copies of the slow allele are also more likely to see their blood pressure and heart disease risk rise if they consume a lot of caffeine.

ALCOHOL FLUSH REACTION

When a person drinks a beer or a glass of wine, the alcohol in it (called ethanol) gets absorbed through the cells of the small intestine and sent into the bloodstream. Alcohol causes intoxication by acting on receptors in our brain that result in relaxation and slower reaction times.

But the alcohol eventually fades from your bloodstream: a single beer doesn't make you feel buzzed for the rest of your life. For that, you can thank your liver.

In your liver, an enzyme called alcohol dehydrogenase (ADH) converts the alcohol into another chemical called acetaldehyde. A second enzyme, aldehyde dehydrogenase 2 (ALDH2), converts the aldehyde into acetic acid. Our kidneys can then take the acetic acid from our blood, and we excrete it in the form of urine.

There's a problem, though: the first enzyme works faster than the second one. If your version of ALDH2 is especially slow, you can end up with a ton of acetaldehyde in your bloodstream that hasn't been metabolized yet. The effects of acetaldehyde can include red, flushed skin, and that's why some people (but not others) look flushed after drinking alcohol. The gene for ALDH2 is located on chromosome 12, and there's a common mutation that slows down its activity. People with that allele have the alcohol flush reaction.

CODEINE METABOLISM

Codeine is an opioid drug that can relieve pain, with the side effects of treating coughs and diarrhea. The enzyme in the liver that metabolizes codeine is CYP2D6. This enzyme actually works on a lot of different drugs—up to 25 percent of all prescription medications. There are at least 160 known variations in the CYP2D6 gene, but most of them don't change its action.

When CYP2D6 acts on codeine, it converts it to the powerful painkiller morphine. There are two rare mutations that can interfere with this process. One makes the enzyme work more slowly, so that the person who takes codeine ends up with only a tiny bit of morphine coursing through their veins. The result: little to no pain relief. Another mutation is even more devastating: it makes the CYP2D6 enzyme work too fast, causing a sudden flood of morphine in the bloodstream. This can result in a standard-sized dose of codeine leading to an overdose of morphine.

For now, most medications are one size fits all: doctors understand what they usually do, in most patients, and they prescribe them accordingly. Some drugs, like antidepressants, work better in some people than others, so if you are trying to manage your depression your doctor may have you try one medication and then, if that one doesn't work, try another.

But ideally, if we can predict what a drug will do based on your genetics, that might let your doctor figure out which drug you need *before* you start trying all of them. When scientists learn more about pharmacogenomics, this personalized approach to medicine will become more common.

ANTIBIOTIC RESISTANCE

How Bacteria Outsmart Us with Evolution

Antibiotics are a critical part of medicine in modern times. Because of these bacteria-killing chemicals, we no longer fear bubonic plague or scarlet fever the way we used to. Minor injuries no longer routinely lead to life-threatening infections. But antibiotics don't kill all bacteria. More and more infections these days are drug resistant, and it's all due to what's going on in the germs' genetics.

The First Antibiotic

Penicillin, the first antibiotic, was not invented; it was discovered. Alexander Fleming, a bacteriologist in London, returned from a vacation in 1928 to find that one of his Petri plates of *Staphylococcus* bacteria had some mold on it. And in a halo around this bit of fungus, no bacteria were growing. It took over a decade of lab work to figure out how to extract and purify the drug from the fungus in great enough quantities to use it in medicine.

The word *antibiotics* means "against life," but the way most people use the word, it means antibacterial drugs. There are also antifungal drugs, antiviral drugs, and antibacterial chemicals that aren't drugs. Their stories are similar, but in this section we're looking at bacteria-killing drugs.

The more than one hundred antibiotics used in medicine come in different classes. All the antibiotics in a class work approximately the same way. Here are some of the major types, including some drugs you may have heard of (or taken!):

- **Beta-lactam antibiotics:** All cells, including bacteria, have a cell membrane around the outside. But some types of bacteria also have a cell wall, a thick barrier made of proteins and sugars that acts as the cell's armor. To build the cell wall, the bacterium has to link chains of proteins together. An enzyme called PBP does this job, grabbing a certain kind of amino acid chain and using it to lock the proteins together. Beta-lactam antibiotics like penicillin are roughly the same shape as the amino acids PBP needs. But when the PBP grabs the antibiotic, the antibiotic gets stuck. That PBP is now out of commission and can't finish building the wall. With enough of the antibiotic around, the bacteria can't build their cell walls, and they can't survive. This enzyme is named after the antibiotic, by the way: PBP stands for *penicillin-binding protein*. Beta-lactam antibiotics include penicillin, amoxicillin, cephalosporins, and carbapenems.
- **Macrolide antibiotics:** Bacteria, like other forms of life, use ribosomes to translate mRNA transcripts into proteins. Macrolide antibiotics attach to bacterial ribosomes, which fortunately are built just a little differently from human ribosomes (so the medicine kills the bacteria, but doesn't kill us). Erythromycin is one example of a macrolide. If you've ever taken a Z-Pak, you've had azithromycin, another macrolide.
- **Quinolone antibiotics:** When bacteria divide, they have to duplicate their DNA (so do we, of course). To do this, they have to separate the two strands of DNA, but this is a tricky job when you remember that DNA is twisted. Next time you happen across a ball of yarn, try grabbing the middle of a strand and pulling it apart. The fibers of the yarn are wrapped around each other, and when you pull them, they'll end up curling even tighter. Bacterial DNA is a continuous circle, so the strands can't unwind on

their own. They have enzymes whose job is to carefully cut and mend the strands of DNA during replication so they don't get too twisted. Quinolone antibiotics stop this enzyme from doing its job. Quinolones include the fluoroquinolone antibiotics levofloxacin and ciprofloxacin (Cipro).

RESISTANCE IS ALL ABOUT BACTERIAL GENES

Each class of antibiotics works on a certain bacterial protein: penicillin binds to PBP, macrolides mess with ribosomes, and quinolones jam up the enzyme that unwinds DNA. That means that bacteria can dodge the effects of these drugs if they can just change how they make these enzymes. In other words, they can benefit if they get just the right mutation in their DNA.

Some of the bacteria that could be killed by beta-lactam antibiotics contain some special genes for enzymes called beta-lactamases. These enzymes can destroy the antibiotic molecule, ripping it apart. Even though penicillin has only been used medically since the 1940s, the fungus that makes it must have been around a lot longer. The beta-lactamase enzymes may be over a billion years old.

Another tactic to dodge the damage of beta-lactam drugs is to tweak PBP so that the antibiotic can't bind as well. This is how MRSA, or methicillin-resistant *Staphylococcus aureus*, accomplishes the methicillin resistance that is its claim to fame.

Macrolide resistance usually comes from a change to the ribosome. But it's not a genetic change; the ribosome's RNA sequence remains the same. Instead, another gene makes a protein called Erm that can add a methyl group to a certain spot on the ribosome. With that small addition, the ribosome is less susceptible to the effects of the antibiotic.

Quinolone resistance can result from mutations that change the DNA unwinding enzyme to make it less susceptible. Another common means of resistance is for the bacterium to acquire efflux proteins. These are pumps that can move antibiotic molecules out of the cell. Efflux pumps can remove many different kinds of chemicals, so if a cell can pump out quinolones, it can often pump out other antibiotics too.

The mutations mentioned here are only some of the sources of antibiotic resistance. Bacteria often accumulate multiple mechanisms of resistance: making several changes to the DNA unwinding enzyme, for example, or having genes for more than one type of beta-lactamase.

SURVIVAL OF THE LEAST SUSCEPTIBLE

The genetic changes that lead to antibiotic resistance don't always occur on the bacterium's main chromosome. Remember how bacteria can transfer tiny circular bits of DNA, called plasmids, to one another? Antibiotic resistance genes often occur on those plasmids.

If you're a bacterium, and someone gifts you a plasmid with an antibiotic resistance gene, that gives you the ability to survive just a little longer. You'll have more opportunities to divide, and of course you're making extra copies of the plasmid so all your children get this helpful gene too. Before long, the antibiotic resistance gene becomes a popular thing: if antibiotics are around, antibiotic-resistant bacteria will flourish while susceptible germs will die.

The same is true of antibiotic resistance genes on the main bacterial chromosome. If you have a gene that gives you resistance, you'll be able to survive longer and have more children. In many cases, antibiotic resistance genes don't give perfect protection: if you're totally swamped with antibiotics, they could still kill you anyway.

That's why, if you get a prescription of antibiotics from your doctor, it probably comes with a sticker warning you that you *must* finish the entire course of antibiotics. The idea there is to keep up a large concentration of antibiotics in your body until all the susceptible bacteria are killed.

But resistance flourishes anyway. If you had a particularly tough bacterium in your body, it might be able to survive even the regular dose of antibiotics. And when you stop taking the antibiotics, that bacterium might even be able to gift its plasmid to other bacteria that come to infect you later. Even our harmless gut bacteria can get in on this gift exchange: after you take antibiotics, you're likely to end up with some antibiotic-resistant bacteria in your gut, and they can give plasmids to other, more harmful bacteria later on.

Because antibiotic-resistant bacteria are more likely to survive over time, the antibiotics represent a **selection pressure** that drives evolution. The process by which susceptible germs die off and resistant ones flourish is called **natural selection**.

To prevent antibiotic resistance from becoming more common, finishing our prescriptions isn't enough. It's also important that doctors only prescribe antibiotics when they are absolutely necessary. That's why your doctor might decline to give you antibiotics for a cold or flu: those are both viral illnesses, and antibiotics don't do anything to kill viruses. Doctors are also advised to test the bacteria that cause an infection to figure out whether they are resistant, and to which antibiotics. That way, your doctor can prescribe the most precise antibiotic that will actually do the job, instead of just guessing at a prescription that might put a selection pressure on all of the bacteria in your body, including the harmless ones.

Antibiotic resistance is also driven by antibiotic use on farms. Animals often get antibiotics in their feed because it helps them grow larger, which is convenient for the farmer: more meat in less time. But this means that farmers are also breeding antibiotic-resistant bacteria alongside their cows or chickens. To tackle this problem, it's best if farmers reduce their dependence on antibiotics, and that they make sure that the ones they do use are not the same antibiotics that are commonly used in human medicine.

QUESTIONS AND ETHICAL QUANDARIES

What You May Not Want to Learn

Your DNA is such an integral part of you that learning about your DNA can give you more information than you really wanted to know. In this section we'll look at some of the questions and concerns that arise when you learn what's in your own genome—or when somebody else finds out.

YOU MIGHT NEED TO REDRAW YOUR FAMILY TREE

Some of the most fascinating things our DNA can tell us include where we come from, and who we are related to. But those same aspects can also be devastating if they reveal that your family tree isn't what you thought.

Alice Collins Plebuch told *The Washington Post* about something unusual that happened when she used a genetic test to find out more about her father's side of the family. Both of her parents were Irish, and she'd met plenty of Irish relatives on her mother's side. Her father was Irish too, and had grown up in an orphanage along with his siblings. But when Plebuch got her results back, they said she was only half Irish; the other half of her DNA reflected Ashkenazi Jewish heritage from Eastern Europe.

GENETICS 101

Plebuch's parents were deceased, so she asked her siblings and cousins to get their DNA tested too. Her siblings shared the half Jewish ancestry, while her cousins on both sides were fully Irish. She was totally unrelated to her cousins on her father's side. She told *The Washington Post* that she spent years confused, disappointed, and determined to figure out what was going on in her family tree. Eventually it turned out that her father was born to a Jewish family but was accidentally swapped at the hospital with a baby from Plebuch's Irish family. Nobody knew, and both of the babies grew up, had children, and went to their graves without knowing that anything unusual had happened.

Plebuch was only able to make the connection when a woman who thought she was Jewish turned out to be half Irish. She was the daughter of the Irish baby that went home with the Jewish family, and a DNA service connected her to Plebuch's Irish cousin.

Or to take a simpler example, a man writing as "George Doe" told the online publication *Vox* that he was surprised to see someone he didn't know turn up as a possible grandfather on a personal genomics site. He asked his parents to take DNA tests as well, to try to figure out how he was related to this mystery person, and it turned out that the person was not a grandfather at all, but a half brother who had been adopted and was looking for his birth family. (A grandfather or a half brother would each share a quarter of their genome with the test taker, hence the confusion.) It turns out George's father had fathered a child that nobody in George's family knew about. George's parents ended up getting a divorce when the truth came out.

Users of DNA tests have unearthed other unwelcome secrets about their families. You may find out that the person you thought was your father turned out not to be your father at all, for example. It's hard to know how often this happens, but scientists' estimates of "nonpaternity events" hover around 2 percent of the population. In 2013, with about a million users in their database, 23andMe reported that seven thousand people had discovered parents or siblings they didn't know about.

PRIVACY CONCERNS

Another big issue with DNA testing is what happens with your data. The personal genomics company that tests your DNA ends up with your sequence or your genotype data on their computers—so what next? Some companies will also store your spit sample in case they want or need to test it again.

All of the major personal genomics companies say they safeguard your data and will never disclose it with your identifying information attached. But that does mean that they may share it under some circumstances. For example, 23andMe shares data with researchers if you have checked the box saying you consent to your data being used for research.

Some companies reserve the right to use your information, including genetic data (but also other information you give them, like the answers to questions about your health) for other purposes, including potentially marketing products to you. And as long as a company holds your data, it's possible that law enforcement could demand that they turn it over, for example, to help solve a crime.

There's also a possibility that personal genomics companies could be hacked and the data stolen. After all, if password and credit card databases can be breached, it's not impossible that DNA data could be taken as well. You can always change your password or get a new credit card, but your DNA is yours forever.

In the US, a law called the Health Insurance Portability and Accountability Act (HIPAA) provides a small amount of protection for medical records: healthcare providers aren't allowed to disclose your records without your permission, and have to take certain steps to keep that data private. But personal genomics companies aren't required to abide by the same law.

There is also a US law called the Genetic Information Nondiscrimination Act (GINA), which prevents health insurance companies from denying coverage or charging you more based on your genetic test results. The law also prohibits employers from making decisions about hiring, firing, or promoting people based on their genes. But there are major loopholes in the law. For example, life insurance, disability insurance, and long-term care insurance aren't included, and can use your information however they like.

YOUR DNA ISN'T JUST YOURS

You share half of your DNA with your biological parents and siblings, and an average of a quarter of your DNA with each of your half-siblings, grandparents, and grandchildren. That means if you learn something about your DNA, your relatives now know something about theirs.

Take the Irish family, for example: once Alice Collins Plebuch discovered that she had Ashkenazi ancestry, that immediately made her siblings question whether they, too, didn't have the heritage they thought. On the other hand, if they hadn't gotten tested as well, Plebuch could be left wondering whether all the siblings had a Jewish ancestor or whether she was the only half-Jewish child—perhaps her mother had had an affair. So even if *you* are fine with learning about your family's mysteries, your other relatives might not be.

The same conundrum applies to disease risk. If you have an allele of the BRCA2 gene that predisposes you to breast cancer, that means there's at least a 50 percent chance that your mother, sisters, or daughters have the allele too. They may not want to know their risk, but if they know yours, they've already learned something about themselves.

GENETICALLY MODIFIED CROPS

Not So Scary After All

Genetic modification is a tool; it's not a bad thing in itself. When you ask people what they think of genetically modified organisms, or GMOs, the response is usually negative. But if you ask people *why* they don't like GMOs or don't want to eat them, the replies often aren't really about GMOs. People tend to say they are concerned about pesticide residue in their food or the effects pesticides have on the environment. Or perhaps they don't like the business tactics of Monsanto, one of the highest-profile companies in the GMO crop business. These are all separate from the issue of whether a crop owes some of its traits to genetic engineering technology.

The truth is that genetic modification in crops is not inherently good or inherently bad. In this section, we'll look at some of the common types of GMOs, and what they really mean for agriculture and for health.

CORN AND SOY ARE OFTEN GENETICALLY MODIFIED

Chances are, you probably ate food made from genetically modified crops today. But that's not because most crops are genetically modified; it's because a few of our most commonly consumed crops are often grown from genetically modified varieties. Here are some of the most commonly grown crops that have genetically modified varieties on the market.

Corn

Most of the corn grown in the US is grown as a grain. (The corn we're talking about is different from the sweet corn you might eat off the cob, slathered with butter.) This corn is the source of cornstarch, corn syrup, and tons of other corn-related products. It also makes up most of the diet that chickens, cows, and other livestock eat. The majority of corn grown this way is genetically modified to stave off insects or to tolerate being sprayed with weed killers, or both.

Soy

Soy is another major component of our food, providing a few protein-containing foods like tofu and the little soy chips in protein bars. It's also the source of more than half of our vegetable oil, and it's also a major component of animal feed. Most soy is genetically modified to resist insects or to tolerate being sprayed with weed killers.

Cotton

Cotton plants make our clothes, but they are food crops too: cotton seed oil is edible. Like corn and soy, cotton plants are often genetically modified to resist insects or to tolerate weed killers.

Potatoes

Potatoes are another commonly grown crop, but most potatoes we eat aren't genetically modified. There is one GMO potato variety that was approved in 2017. Its naturally occurring enzyme for browning (which happens when the potato is bruised) no longer works, so the potato doesn't get black spots. This potato also produces less acrylamide, a chemical that is formed when food is fried or cooked at high temperatures.

Papayas

Most papayas on the market today are genetically engineered to be resistant to the papaya ringspot virus. The GMO papaya may have saved the papaya industry; before it came along, the virus devastated papaya crops.

One You Won't See

One crop that's *not* on this list: tomatoes. A genetically modified tomato was among the very first GMO crops to be approved, so you may see news articles on GMO crops illustrated with pictures of tomatoes. But the GMO tomato didn't work out commercially. Its makers stopped growing it shortly after it was introduced, and it's been off the market since 1997.

INSECT-RESISTANT CROPS

Crops can be genetically engineered to be resistant to insects, like caterpillars, that would otherwise eat them. The first commercial success, and currently very common in crops, uses a toxin from *Bacillus thuringiensis*.

B. thuringiensis, often nicknamed Bt, is a bacterium. Some strains of Bt can produce a toxin that forms sharp, pointy crystals inside the guts of caterpillars. Since the 1920s, farmers and gardeners have used spores of the bacteria as a natural insecticide. If you buy Thuricide at the garden store, for example, that's a solution of bacterial spores meant to spray on your plants. The toxin only affects certain insects, so it's considered a natural and safe pesticide and is allowed in organic farming.

The first genetically engineered Bt corn was approved in 1996. It makes a toxin that targets a common corn pest called the European corn borer. To make this type of corn, scientists copied the gene from Bt bacteria that produces the toxin and inserted that gene into the genome of the corn plant itself. The effect is similar to constantly spraying Bt bacteria all over corn, except without the spraying. The plant itself makes its own insecticide, and as a result, farmers can spray less chemical insecticide.

HERBICIDE-TOLERANT CROPS

The other major category of GMO crops includes herbicide-resistant plants. Herbicides are weed killers.

When the corn or soy in a field is resistant to a certain herbicide, you can spray the herbicide over the whole field and expect that it will only kill weeds, while leaving the crop unharmed. While insect-resistant crops tend to let farmers spray less insecticide, herbicide-resistant crops tend to result in more spraying of herbicides. That's why it's best to be wary of claims that GMO crops are all good or all bad; it depends on what kind of crop it is and how it is cultivated.

The first herbicide-tolerant GMO crops were "Roundup Ready" soybeans, created to tolerate the herbicide glyphosate, sold by Monsanto under the brand name Roundup. Glyphosate kills plants by interfering with an enzyme that makes amino acids they need to live. So scientists inserted a gene into soybeans so they could make a bacterial version of that enzyme, instead of the usual plant version. The bacterial version can still work even in the presence of glyphosate.

The widespread use of these crops meant that the use of Roundup became more common. Just as bacteria can develop resistance to antibiotics, weeds can also develop resistance to weed killers. Glyphosate is a common pesticide; you can buy it in garden stores for your own use, no GMO plants required. Thanks to GMO soybeans, though, its use became even more common.

Over time, seed companies like Monsanto and Dow have come up with other combinations of herbicides and herbicide-tolerant plants to give farmers more options. Still, resistant weeds are a constant issue.

GENETIC ENGINEERING TOOLS

Tinkering with Bacteria, and More

Over the years, scientists have learned about some weird and cool things you can do with DNA. These findings have helped us to understand what goes on naturally in our cells (and elsewhere in nature, such as bacterial cells), but they also open up the possibility of using enzymes, viruses, and more as tools.

The techniques we'll mention in this section are some of the workhorses of genetic engineering. They make many things possible, from medical advances to paternity tests to vegetarian cheese.

POLYMERASE CHAIN REACTION (PCR)

If you want to do anything with DNA, you have to make sure you have enough of it to work with. A technique called the polymerase chain reaction (PCR) can amplify a tiny tissue sample from a crime scene into enough DNA for analysis. Personal genomics companies also use PCR to get usable DNA out of the small number of cells floating in your spit sample. PCR can also be used to separate out the parts of DNA you want to work with from the rest of the genome that you don't really care about.

PCR uses DNA polymerase, the same enzyme that copies DNA in our cells when it's time for them to divide (see the "Making More Cells" section). But instead of the human version of that enzyme, PCR uses a version of that enzyme, called Taq polymerase, from a bacterium called *Thermophilus aquaticus*. This tiny creature lives in very hot water, including the geysers in Yellowstone National Park. Its DNA polymerase can withstand extreme heat.

That's convenient, because when we heat DNA, the two strands of the double helix are no longer strongly attracted to each other. So a PCR machine heats a sample of DNA until the two strands come apart. Then it lowers the temperature back to normal, and the Taq polymerase gets to work filling in the missing nucleotides on each lonely strand. Now we have two pieces of DNA. Raise the temperature again, and the same thing happens. In each cycle, the amount of DNA doubles, so it only takes ten cycles to get 1,024 copies of DNA.

This polymerase needs a double-stranded bit of DNA to start from, though. So when you begin your PCR experiment, you can add primers—short segments of single-stranded DNA—that match the sequence at the beginning and end of the spot you want to amplify.

RESTRICTION ENZYMES AND LIGASE

If you want to cut and paste a gene from one place to another, restriction enzymes are a great way to do that. These are enzymes naturally made by bacteria, and each one has a specific sequence that it recognizes and then cuts unevenly. You can use this enzyme to cut a gene of interest out of any old DNA, and then use the same enzyme to cut a plasmid (a tiny circular piece of DNA that bacteria can carry). Because the enzyme recognizes a certain sequence, and cuts it unevenly, each of the places that was cut has a length of unpaired DNA with a certain sequence. And it just so happens that all of the unevenly cut ends are able to match with each other. That means you can take the gene you just cut and insert it into the plasmid.

Once you have a plasmid containing a gene you'd like to work with, you can introduce it into bacteria. (There are several ways to do this, including lightly electrocuting the bacterial cells to create temporary holes in their membrane. Bacteria are tough enough that they don't mind.) Then, when the bacteria replicate, your plasmid will replicate as well.

Yeast can be convinced to take up plasmids as well. You can also make an extra-large plasmid called a bacterial artificial chromosome (or, in yeast, a yeast artificial chromosome) and count on the bacteria or yeast to copy that chromosome as long as they live and divide.

Inserting a gene into bacteria is a great way to make a lot of something. There are vats of bacteria that work all day long expressing the human gene for insulin; if you have diabetes and inject insulin, this is where it comes from. This is also where most rennet today comes from. Rennet is an enzyme traditionally harvested from the stomach linings of baby cows that have been butchered. It's necessary to make certain types of cheese. But there are now bacteria with the rennet gene inserted, so no baby cow stomach linings are needed to make cheese anymore. Check the ingredient label next time you buy cheese; if it contains rennet or enzymes but is labeled as vegetarian (as in "vegetarian rennet") then you know it came from engineered bacteria rather than from slaughtered cows.

GEL ELECTROPHORESIS

If you have a soup of DNA fragments, you may want to know what that soup contains. Instead of sequencing or genotyping all the fragments, sometimes all you need is a rundown of how big each fragment is.

For example, if you have some DNA from a crime scene, you can use PCR to amplify the part of the DNA that tends to contain many small repeats (short tandem repeats, or STRs). Everybody has a slightly different number of these repeats; maybe my STR segments are long and yours are short. Years ago, DNA fingerprinting was more often done with restriction enzymes; not everybody has the same number of recognition sequences for a given restriction enzyme, so if you cut up two different people's genomes with the enzyme you could tell who was who based on how big the segments are.

The gel is just what it sounds like—a slab of material that looks and feels a little bit like Jell-O. If you could shrink down to the size of a DNA molecule, you would see the gel as a network of chains of sugars called agarose. A tiny piece of DNA could easily slip between the strands, like how a mouse can disappear into a brush pile. But a larger piece of DNA would have a harder time; if a cat were chasing that mouse, the cat might be able to get into the brush pile but wouldn't be able to move through it as quickly as the mouse.

To get the DNA to move, a scientist hooks up electrodes that put a positive charge at one end of the gel and a negative charge at the other. DNA is negatively charged, so it wants to move toward the end with the positive electrode. After a while, perhaps an hour or two depending on the experiment, the DNA shows up as stripes, or bands, in the gel. DNA is naturally clear, but you can dye it after the fact, or you can use DNA that has already had a dye or label attached.

CRISPR AND GENE EDITING

Cut and Paste

There are many ways to cut and paste DNA, including several that scientists have been using for decades. But one new technique, known as CRISPR/Cas9, can do the job easily and more precisely than ever before.

Previous techniques didn't give scientists a lot of control over the process. You could cut a gene out of a chromosome, as long as the sequence had a recognition site for a restriction enzyme in the right place. And if you wanted to paste that gene into another organism, you would need to use a plasmid, or a virus, or some other vector that you can't always depend on to put the gene in a certain place. Eventually scientists figured out how to create custom proteins, like zinc finger nucleases, to target specific sequences. But those proteins were expensive to make.

But with a CRISPR/Cas9 technique developed just a few years ago, you can actually tell the CRISPR machinery where in a chromosome you would like to see a change and exactly what change you would like it to be. Jennifer Doudna, one of the people who discovered the technique, has said that it's like having a scalpel when the previous tools were all sledgehammers.

WHERE CRISPR/CAS9 CAME FROM

Like most genetic engineering tools, CRISPR/Cas9 is not purely a human invention, but is a clever way to use something that was already found in nature.

CRISPR is part of a bacterial immune system. That's right: germs can get sick too. So bacteria and archaea keep track of invaders they have seen before, the better to recognize and destroy them next time.

In bacteria, the invaders are typically viruses. Bacteria fight them by chopping their DNA (or RNA) into pieces. The bacteria can copy a snippet of the virus's sequence into their own chromosome, making a scrapbook of all the invaders they have defeated.

The snippets in the scrapbook are separated by special repeats of DNA. Scientists didn't realize at first why they were finding sections of bacterial DNA with repeats, then a unique sequence, then more repeats, and so on. They named this odd configuration "clustered regularly interspaced short palindromic repeats." For short: CRISPR.

Bacteria that had CRISPR sequences also had genes that made DNA-binding proteins, and these are the Cas, or CRISPR-associated genes.

HOW CRISPR AND CAS9 EDIT DNA

In the wild, the bacterium makes an RNA copy of one or more of its scrapbooked sequences from the CRISPR array, and a protein called Cas9 holds onto it. This is the guide RNA: the thing that Cas9 is programmed to search and destroy. When it finds a piece of DNA floating around in the cell, it cuts that DNA. Mission accomplished.

Scientists figured out that they can get Cas9 to cut any spot in any genome, so long as they give it the right guide RNA so it knows what to look for. That makes this a programmable enzyme.

Cutting DNA sounds like bad news, but don't forget we have systems for repairing damaged DNA. If Cas9 just cuts, the cell can usually figure out how to put the remaining DNA back together. Or scientists can provide an extra piece of DNA for the repair enzymes to insert.

And here's the best part: this system can be used not just in bacteria but in any type of cell, including human cells.

HOW SCIENTISTS ARE USING CRISPR/CAS9

An enzyme that can find a DNA sequence, cut it, and optionally insert any sequence that you want, is a very powerful tool. It's also a very cheap and easy method to use, so you can buy at-home kits for $200 to edit any genes you like—they're intended for bacteria, though, and not your own cells.

CRISPR will likely be the tool of choice for genetically modifying crops in the future. It's already been used to create a mushroom that doesn't brown when cut. The technique works in animals as well as people, so one group of scientists used CRISPR to create leaner pigs whose meat is lower in fat. Another group worked in cattle, snipping out a gene that's necessary for horns to develop. It's possible to breed hornless cattle, but it would take decades of careful mating. CRISPR does the job in just one generation.

Another use of CRISPR in livestock is relevant to human health. Decades ago, scientists explored the idea of transplanting organs from pigs to people, because there aren't enough donated human organs to go around. But a pig's genome contains retroviruses that could be dangerous to a human host. A team of researchers recently managed to remove all sixty-two of these problematic elements from a pig's genome, opening up the possibility of animal-to-human transplants once again.

HOPES AND FEARS FOR THE FUTURE

But porcine transplants aren't likely any time soon; we still don't know if they would be safe. The same is true for many other potential uses of CRISPR.

Another exciting but problematic project is a "gene drive" for mosquitoes. Scientists can create mosquitoes in the lab that resist infection with malaria. And they can create them with the genes for an entire CRISPR-based editing system right in the DNA. That means that a lab-grown mosquito could breed with a wild mosquito, and the resulting offspring will briefly have one antimalaria gene. But then that gene can use its CRISPR machinery to copy and paste itself into the mosquito's other chromosome. It's a way of ensuring that a gene propagates through a population a lot faster than it otherwise would.

But this means we would be irreversibly changing the genetics of an entire population of mosquitoes, maybe eventually an entire species. It's designed as a runaway technology, so if anything goes wrong and we decide this is a gene that *shouldn't* spread, there's no way to fix it.

Perhaps the scariest of all is the idea of using CRISPR on humans. This is also a cause for hope, though: perhaps we could edit diseases out of our genomes, or alter cancer cells' genomes so they stop acting cancerous.

CLONING AND DE-EXTINCTION

Cloning Means Never Having to Say Goodbye

To a genetics nerd, there has probably been no movie more thrilling than the 1993 *Jurassic Park*. Not only were there dinosaurs, and not only were the dinosaurs more realistic than movies typically made them at the time, but the whole story was based on some fairly plausible genomic science.

To be clear, it's not *super* plausible; we're not going to see supersized velociraptors running around any time soon. But the idea of bringing an extinct creature back to life is something scientists are still thinking about. Is it ethical? Is it possible? If we can sort those questions out, what's the best way to do it?

CLONING

In *Jurassic Park*, scientists started with dinosaur DNA found in the blood in the belly of a mosquito preserved in amber. They injected this DNA (uh, somehow) into the egg of a modern-day emu or ostrich, creating an embryo that was able to divide and develop into a hatchling dinosaur.

We can't yet clone extinct animals this way, but *living* animals have been cloned. In 1996, just a few years after *Jurassic Park*, a team of biologists in Scotland managed to clone a sheep.

To do this, they took an egg cell from one adult female sheep and removed its nucleus. They replaced it with the nucleus from another cell, taken from the mammary gland of a second sheep. Since the nucleus they transferred was from a somatic (body) cell and contained a complete diploid set of chromosomes, no sperm was needed. This technique is called **somatic cell nuclear transfer**. A mild electric shock triggers the cell to start dividing.

The scientists implanted the resulting embryo into the uterus of a third sheep who would carry the pregnancy. As a result, Dolly, the world's first cloned sheep, had three mothers and no father.

Are Clones Healthy?

Dolly died at age six, younger than the typical lifespan for her breed, and at first the scientists were worried that clones weren't as healthy as naturally conceived sheep. But other cloned sheep lived longer lives than Dolly did, and recent studies showed that the arthritis in her knees—at first thought to be a sign of premature aging—was actually in the normal range for a six-year-old sheep.

Since then, other animals have been cloned—but nobody has attempted a human. The first cloned horse, Prometea, was born in 2003. The rules around Thoroughbred horse racing don't permit clones, but other valuable horses have been cloned. Polo star Adolfo Cambiaso won a match riding six clones of his deceased favorite horse, Cuartetera.

Cloned animals aren't always identical to their parent, though. People who have cloned their pets (for prices ranging from $25,000 to $100,000) report that the resulting animals often have a different personality and sometimes slightly different markings on their fur. Remember, DNA doesn't determine *everything* about you, and that's true of animals too.

BRINGING BACK EXTINCT SPECIES

Our cells do a lot of work to keep DNA in good shape. So when cells die, the DNA can become damaged, and nobody is there to patch it back up.

Once DNA has been sitting around for thousands or millions of years in a fossil, it's somewhere between fragmented and obliterated. That's a big part of why dinosaur cloning is still more fantasy than science fiction.

But some species died off more recently. The Neanderthal Genome Project was able to recover enough DNA from a bone that was over fifty thousand years old to deduce that Neanderthals were 99.7 percent similar to modern *Homo sapiens*, and that today's Europeans may have inherited up to 4 percent of their DNA through Neanderthal ancestors.

In 2000, the world's last Pyrenean ibex died in a nature preserve in Spain. Scientists had already saved a sample of tissue from her ear, and in 2003 they implanted the nuclei of some of her cells into egg cells from goats. Only one of the embryos was born, and it died shortly after birth from a lung defect. They could try again, but even if they successfully clone this female ibex, there are no males of her species to breed with. There is, however, a related species called the Southeastern Spanish ibex, so a hybrid would be possible.

Another animal that became extinct recently is the quagga, a subspecies of zebra that had stripes on the front half of its body but was a plain brown in back. Since it's the same species as modern zebras, scientists hypothesized that all the genes necessary to make a quagga already exist in living zebras; the challenge is just to bring them together in one animal. The Quagga Project is an attempt to recreate the quagga—or at least something that *looks* like a quagga—through selective breeding.

The woolly mammoth is another animal that scientists, including Harvard geneticist George Church, are considering bringing back. No intact cells of woolly mammoths exist, so cloning isn't possible. But mammoths only died out a few thousand years ago, and there are some extremely well-preserved specimens. Some mammoths lived in places that were cold enough that their bodies froze shortly after death, and they have remained frozen ever since. A baby mammoth was found in 2013 with hair, muscle tissue, and liquid blood still recognizable.

Church's team believes they can recreate a mammoth by analyzing mammoth DNA and determining where it differs from modern elephants—for example, finding the genes that cause mammoths to grow fur. Then they plan to use gene editing techniques like CRISPR to create an elephant embryo with mammoth DNA in just the right places, and grow it in an artificial womb. It's an ambitious project, and only time will tell if it's truly possible.

BABIES OF THE FUTURE

Reproductive Tech and "Designer" Babies

If the scary movie model for de-extinction is *Jurassic Park*, then the corresponding movie for designer babies is 1997's *Gattaca*.

There aren't currently any "designer babies," but that's the term that gets thrown around when any genetic or genomic technology emerges that could potentially be used to let parents dictate or choose traits of their offspring-to-be.

SCREENING EMBRYOS

In *Gattaca*, a dystopian movie about how widespread DNA sequencing might change our society, parents don't edit their children's genomes. They just get the opportunity to choose between a variety of embryos produced with their own eggs and sperm, but whose genomes have been sequenced for a peek into their future.

This is fiction, but something similar is available for parents who carry alleles for certain genetic diseases. The technique is called **preimplantation genetic diagnosis (PGD)**.

Here's how it works: when an embryo has only eight cells, you can remove one cell to test its DNA, and the embryo will still develop as normal. This type of screening was first used in 1990 to select female embryos for a couple who was at risk of passing down X-linked genetic disorders to their male offspring. Since then, PGD has been used to test for specific genes including those for Huntington's disease and cystic fibrosis. Technicians will test multiple embryos for the disease and only implant those that don't have the affected allele.

Some European countries prohibit PGD, including Austria, Germany, Ireland, and Switzerland. Other countries, including France, Greece, and the United Kingdom, allow it for medical reasons only. The United States does not regulate the technique.

GENE EDITING

Another ethical dilemma is the idea of human germline editing. The germline means any cells that will eventually become a person. It includes eggs and sperm and also embryos themselves. If a parent is doomed to give their child a disease-causing gene, CRISPR could be used to fix that gene at the embryo stage. This becomes an ethical minefield: if it's okay to fix a lethal disease, is it also okay to edit an embryo to change its risk for less serious conditions? What about edits that would make a resulting child taller or better at sports?

So far the only published CRISPR experiment on a human embryo was done with cells that did not develop into a baby. The experiment was a mixed success: the scientists successfully made edits to the cell, but in some embryos the edit did not take, and in others, the embryo ended up with unwanted random mutations.

THE ETHICS OF DESIGNING YOUR BABY

Screening for genetic diseases has been around longer than the ability to edit embryos. Even before PGD, pregnant people have been able to get tests, such as amniocentesis, to check on their developing offspring's DNA. Amniocentesis involves taking a sample of the amniotic fluid surrounding a fetus. It's a risky procedure, since it carries a small risk of miscarriage. But it can reveal whether the fetus has a normal number of chromosomes, or if it has Down syndrome or another condition related to the number of chromosomes.

The bigger ethical dilemma surrounding genetic and genomic testing for babies is whether it's okay to change or select an embryo based on genes that *don't* relate to lethal or serious genetic disorders. For example, parents might want to have a baby that is smart, or tall, or attractive, or good at sports—or all of the above.

We don't have to worry about this kind of genetic engineering yet, simply because we don't know all of the genes that go into making somebody smart or tall or attractive or good at sports. Hundreds of genes influence height, for example. Meanwhile, we have no idea how many genes influence intelligence or attractiveness, in part because nobody can agree on what intelligence or attractiveness really are.

But gene editing techniques are becoming more advanced, and sequencing is getting cheaper and cheaper. Some scientists and bioethicists treat embryonic gene editing as an issue we must be prepared for, saying it's likely that one of these days somebody will try it.

In *Gattaca*, people whose genomes had been screened before birth were considered to be a higher, more privileged class of people. Ethicists worry that something along those lines really could happen, regardless of how sound the science is behind it. Not long ago, in the early 1900s, some scientists were calling for the science of genetics to be used to encourage people with "good" genes to have offspring, and for people who were poor, disabled, or had criminal records to be sterilized.

We don't need a return to those days. But the science is moving fast, and only time will tell what will happen with reproductive technology in the age of gene editing and cheap sequencing.

GLOSSARY

Adaptation

The phenomenon of living things becoming more suited to their environment, due to natural selection occurring over many generations.

Allele

One of two or more possibilities for a gene or other DNA sequence. For example, the genes for A and B blood proteins, in humans, are two of the alleles possible at the ABO blood type locus.

Alternative splicing

A process that allows the same gene to code for multiple proteins or products, depending on how its mRNA is spliced.

Amino acid

The building blocks of protein, amino acids are small molecules with a nitrogen-containing amino group and a carboxylic acid group.

Antibiotics

Medicines that kill or stop the growth of microorganisms, especially bacteria.

Apoptosis

Cellular self-destruction, often performed when the cell (such as a cell of the human body) is defective or potentially cancerous.

Atom

The smallest possible unit of a chemical element. Multiple atoms, bonded together, can make molecules.

Autosome

One of the paired, numbered chromosomes (in humans, chromosomes 1 through 22).

Bacteria

Single-celled creatures without a nucleus, categorized in the domain Bacteria. Some are pathogens, like *Salmonella*, while others live harmlessly in our bodies and our environment.

Bacteriophage

A virus that infects bacteria.

Base pairing

The nitrogenous base portion of a nucleotide matching up to another. Adenine pairs with either thymine or uracil; cytosine pairs with guanine.

Cell

The smallest unit of a living thing, enclosed by a membrane. Some creatures only have one cell (such as bacteria), while others are made of many cells (such as humans).

Central dogma

A concept in biology (not actually a dogma) that describes how the information in RNA is used to build proteins, but not the reverse.

Centromere

The part of the chromosome where microtubules attach during cell division. It is often located near the center of the chromosome.

Chromatid

Each of the strands of DNA that results from DNA replication. During some stages of cell division, multiple chromatids may remain joined together in a single chromosome.

Chromatin

The combination of DNA and histone proteins.

Chromosome

A strand of DNA, with its associated histone proteins, especially when it is condensed enough to be visible under a light microscope.

Cloning

The process of creating a new individual whose nuclear DNA is identical to another individual. In plants, this can happen by cutting off a piece and rooting it in new soil. In animals, it can only happen through manipulation of cells in the lab. (See somatic cell nuclear transfer.)

Codon

Three letters of RNA that match to a known tRNA/amino acid complex.

Complementary strand

The DNA or RNA strand opposite the strand we are talking about.

Condensed

Chromatin that is bundled tightly together and usually not expressed.

Conserved

Refers to a DNA sequence that has changed very little over time. Conserved genes or regions are usually those that are important to a living thing's ability to survive and function.

Covalent bond

Chemically, a bond between two atoms that results from sharing electrons.

CRISPR/Cas9

A gene editing system originally discovered as a sort of bacterial immune system. Scientists can provide an RNA sequence to direct a CRISPR/Cas9 system to cut DNA in a specific place.

Cytoplasm

The portion of a cell that is not inside any of its organelles.

Deletion mutation

A change in the DNA in which one or more nucleotides are lost.

Diploid

Containing two copies of each chromosome. Most human cells are diploid.

DNA

Deoxyribonucleic acid, the genetic material in living cells.

DNA polymerase

An enzyme that can attach nucleotides to each other, forming a DNA strand.

Dominant

A gene whose effects can still be observed even in the presence of another allele for the same gene.

Double helix

The shape of DNA: two strands that twist around each other.

Enhancer

A DNA sequence, located some distance away from a gene, that can cause that gene to be more likely to be transcribed. A transcription factor that binds to the enhancer becomes part of the transcription complex.

Enzyme

A protein that catalyzes a chemical reaction.

Epigenetics

The study of heritable changes in gene expression.

Epistasis

Interactions between multiple genes that control a single phenotype.

Eukaryote

Living things that have a nucleus in their cell. Animals, plants, and fungi are all eukaryotes, but bacteria are not.

Exome

The collection of all DNA in a genome that is transcribed and translated.

Exon

A section of mRNA (or in the DNA that produces it) that is kept after splicing.

Express

To use the instructions from a gene to make the gene's product.

Folding of proteins

A change in shape that proteins undergo during or after translation. Each protein begins as a strand of amino acids (or several such strands), but after folding, the protein has a specific shape and function.

Frameshift mutation

A change in a gene that disrupts the sequence of three-letter codons, changing their meaning.

Gender

The personal or cultural experience of being male, female, or neither, as distinct from biological sex.

Gene

The sequence of DNA that provides instructions for building a gene product such as a protein or a functional RNA.

Gene editing or genome editing

Laboratory techniques for cutting and inserting DNA.

Genetic material

The molecules that carry instructions for producing traits in a living thing and that are passed on from parent to offspring. This term predated our understanding of DNA. We now know that DNA is the genetic material.

Genetics

The study of genes and how they are inherited.

Genome

All of the DNA in a living thing.

Genomics

The study of whole genomes.

Genotype

The sequence of DNA at particular locations (loci) of interest. Contrast with phenotype.

Genotyping

Testing a sample of DNA to determine which genotypes it contains at specific locations.

GMO

A genetically modified organism, specifically one whose DNA was manipulated in the lab to add a particular gene.

Haplogroup

A group of living things that have all inherited the same collection of DNA sequences from a recent common ancestor.

Haploid

Containing only one of each chromosome, rather than pairs.

Haplotype

A collection of DNA sequences that were inherited together and that categorize living things into a haplogroup.

Helicase

An enzyme that unwinds DNA during replication.

Heredity

The study of how traits are passed from parents to offspring.

Heterozygous

Having two different alleles of the same gene or locus.

Histone

One of the proteins that DNA is wrapped around.

Homozygous

Having two identical alleles of the same gene or locus.

Hybrid

An individual whose parents were of different, purebred varieties. For example, a cockapoo is a hybrid between a cocker spaniel and a poodle.

Hydrogen bond

An attraction between molecules, often between a positive hydrogen atom on one and a negatively charged oxygen or nitrogen on the other. Hydrogen bonds are what hold the base pairs, and thus the two strands of DNA, together.

Identity by descent (IBD)

The proportion of genetic material shared by descent between two individuals. By this measure, a person shares 50 percent of their genes with each parent.

Identity by state

The proportion of measured alleles that are identical between two individuals. Typically a person is 85 percent identical by state to each parent.

Imprinting

Marking of the DNA, typically by methylation, in the sperm or egg cell.

Insertion mutation

A change in the DNA that results in the gain of one or more nucleotides.

Intron

A portion of an mRNA that is removed in the splicing process.

Karyotype

A photograph of the condensed chromosomes in a single cell, providing a way to see if any chromosomes are duplicated, missing, or have certain unusual changes such as large translocations.

Locus

A location within the DNA sequence of an individual.

Lysogenic cycle

A way viruses can reproduce, by inserting DNA into the host's genome and waiting for the host to reproduce.

Lytic cycle

A way viruses can reproduce, by encouraging the host cell to replicate the virus and then burst open.

Meiosis

A type of cell division that results in a diploid parent creating four haploid daughter cells.

Methylation

An alteration to DNA, used for imprinting or other epigenetic purposes. Methylated DNA is typically not transcribed.

Microbiome

The population of microbes (especially bacteria but also including archaea, fungi, protists, and other microscopic creatures) that live in a particular place.

miRNA

A small ("micro") RNA that folds back on itself and serves to silence gene expression.

Mismatch repair

A system of repairing DNA when one strand contains the correct nucleotide and the other, newly synthesized, strand does not.

Mitochondrion (plural: mitochondria)

An organelle in eukaryotic cells that metabolizes fuel and oxygen into ATP.

Mitosis

A form of cell division that results in daughter cells that are genetically identical to the parent cell.

Molecule

A combination of atoms held together by covalent bonds.

Monosomy

The condition of having only one of a particular chromosome instead of a pair.

mRNA

A strand of RNA that acts as a "messenger" between the nucleus where transcription occurs, and the cytoplasm where translation occurs.

Mutation

A change in DNA.

Natural selection

A process by which mutations or gene variants become more common in a population because they provide an advantage in survival or reproduction.

Neanderthal
An extinct type of human that lived in what is now Europe.

Nitrogenous base
The portion of a nucleotide that contains nitrogen and is chemically basic. Adenine, thymine, uracil, guanine, and cytosine are nitrogenous bases.

Noncoding DNA
DNA that does not carry instructions for making a protein or functional RNA.

Nondisjunction
An error in cell division that occurs when a pair of chromosomes does not separate properly in meiosis I. As a result, one gamete may have two copies of the chromosome, while another of the gametes from that division has none.

Nucleotide
One of the building blocks of a DNA strand, consisting of a sugar, a phosphate, and a nitrogenous base.

Nucleus
The compartment of a eukaryotic cell that carries the cell's DNA (not counting DNA in other organelles, such as mitochondria).

Okazaki fragment
A short length of DNA created on the lagging strand during replication.

Organelle
A compartment or structure within a cell, especially those (like the nucleus and mitochondria) enclosed by a membrane.

Ovulation
The process of releasing a mature egg cell from the ovary, which occurs while that cell is undergoing meiosis.

PCR

Polymerase chain reaction, a laboratory procedure that copies a portion of DNA over and over by heating and cooling it in the presence of a heat-stable DNA polymerase and a pair of primers of known sequence.

Phenotype

The traits or appearance of a living thing.

Phylogenetics or phylogenomics

The study of family trees as determined by the relationship between individuals' or species' genes or genomes.

Plasmid

A small circular section of double-stranded DNA carried by bacteria, outside of their main chromosome.

Pollination

The process of fertilizing a plant's egg cell with gametes contained in pollen.

Polymer

A large molecule made of repeating segments of a building block. DNA is a polymer of nucleotides, and proteins are polymers of amino acids.

Polyploid

Containing many complete sets of chromosomes.

Preimplantation genetic diagnosis

A reproductive technology procedure in which one cell of an embryo is genetically tested, to allow the parents to choose whether or not to implant the embryo.

Prokaryote

A living thing without a nucleus in its cells. Bacteria and archaea are prokaryotes.

Promoter

A section of DNA at the start of a gene. RNA polymerase and transcription factors must attach to the promoter to begin the process of transcription.

Proofreading
A form of DNA repair that occurs during DNA synthesis. The DNA polymerase typically cannot move forward until it removes the wrong nucleotide and replaces it with the correct one.

Protein
A polymer of amino acids, created according to the instructions contained in RNA.

Punnett square
A chart showing the possible combinations of alleles of interest from two parents. Often used to predict the offspring of a hybrid (see the section titled "Dominant and Recessive").

Recessive
A gene whose effects cannot observed in the presence of another allele for the same gene.

Regulatory elements
Sections of DNA that can affect whether or not a gene is transcribed. Promoters, enhancers, and silencers are all regulatory elements.

Replication
DNA replication is the process of separating the two strands of DNA and adding nucleotides to each strand, resulting in two double-stranded lengths of DNA where previously there was one.

Replication bubble
A structure that forms temporarily as the two strands of DNA are separated during replication.

Replication fork
The Y-shaped formation of DNA at each end of the replication bubble.

Retrotransposon
A section of DNA that can copy itself for insertion into another

part of the DNA. These may derive from retroviruses that embedded themselves into the genome in the distant past. Retrotransposons make up at least 27 percent of the human genome.

Retrovirus
A virus that contains an RNA genome and codes for a reverse transcriptase that can synthesize DNA from an RNA template.

Reverse transcriptase
An enzyme, found in retroviruses, that can synthesize DNA from an RNA template.

Ribosome
A complex of proteins and RNA that translates RNA instructions into amino acid sequences to synthesize proteins.

RNA polymerase
An enzyme that can synthesize RNA from a DNA template.

RNAi
RNA interference, a process for silencing genes that have been transcribed but not yet translated.

Selection pressure
A factor in the environment that favors individuals with a certain genotype over those with another.

Sequencing
The process of determining the sequence of a given sample of DNA.

Sex
The state of being male or female, as determined by chromosomes or by anatomy.

Sex chromosome
The chromosomes that determine chromosomal sex in some species, including humans. (Human females have two X chromosomes; males have an X and a Y.)

siRNA

A small interfering RNA, a double-stranded RNA molecule that can silence gene expression (see RNAi).

SNP

Abbreviation for *single nucleotide polymorphism*, pronounced "snip." A genetic variant, involving only one nucleotide, that is shared with at least 1 percent of the population.

snRNP

Abbreviation for *small nuclear ribonucleoprotein* and pronounced "snurp," one of the RNA-protein complexes that carries out splicing on mRNAs.

SNV

Abbreviation for *single nucleotide variant*. A genetic variant involving only one nucleotide.

Somatic cell nuclear transfer

A cloning process, performed in the laboratory, that involves implanting the nucleus of one cell into another cell.

Spermatogenesis

The process of creating sperm cells through meiosis.

Splicing

The process of removing introns from an mRNA.

Start codon

Three letters of RNA code that indicate where translation should start. These three letters are usually AUG, and also indicate that the first amino acid should be a methionine.

Stop codon

Three letters of RNA code that indicate where translation should stop. These are typically UAA, UAG, and UGA.

Structure

The physical shape of a molecule.

Telomere

The repeating sequence at the ends of a chromosome.

Trait

An observable property of a living thing.

Transcription

The process of creating an RNA copy of a section of DNA. It is the first step in gene expression.

Transcription complex

A collection of proteins, including RNA polymerase, required to begin transcription of a gene.

Transcription factor

One of the proteins needed for the transcription complex.

Translation

The process of synthesizing protein using the instructions from RNA.

Translocation

A rare phenomenon in which a section of DNA is removed from one chromosome and attached elsewhere, usually on another chromosome.

Trisomy

The condition of having three of the same chromosome. Down syndrome results from a trisomy of chromosome 21.

tRNA

Short for *transfer RNA*, the RNA that matches a codon to an amino acid.

Virus

A package of DNA or RNA, encased in protein, that is capable of infecting a cell and causing that cell to reproduce the virus.

INDEX

ABOUT THE AUTHOR

Beth Skwarecki is the health editor at Lifehacker.com. She has previously worked as a freelance health and science writer: her work has been featured on *Medscape, Performance Menu, Public Health Perspectives, Bitch* magazine, *The Pittsburgh Post-Gazette, Science,* and *Scientific American*. She is the author of *Outbreak! 50 Tales of Epidemics That Terrorized the World*.